OFFICIAL SQA PAST PAPERS WITH ANSWERS

## ADVANCED HIGHER

# CHEMISTRY
## 2008-2012

First exam published in 2008.
Published by Bright Red Publishing Ltd, 6 Stafford Street, Edinburgh EH3 7AU
tel: 0131 220 5804 fax: 0131 220 6710 info@brightredpublishing.co.uk www.brightredpublishing.co.uk

ISBN 978-1-84948-301-8

A CIP Catalogue record for this book is available from the British Library.

Bright Red Publishing is grateful to the copyright holders, as credited on the final page of the Question Section, for permission to use their material. Every effort has been made to trace the copyright holders and to obtain their permission for the use of copyright material. Bright Red Publishing will be happy to receive information allowing us to rectify any error or omission in future editions.

ADVANCED HIGHER

2008

[BLANK PAGE]

# X012/701

NATIONAL
QUALIFICATIONS
2008

FRIDAY, 30 MAY
9.00 AM – 11.30 AM

# CHEMISTRY
## ADVANCED HIGHER

Reference may be made to the Chemistry Higher and Advanced Higher Data Booklet.

**SECTION A — 40 marks**

Instructions for completion of **SECTION A** are given on page two.

For this section of the examination you must use an **HB pencil**.

**SECTION B — 60 marks**

All questions should be attempted.

**Answers must be written clearly and legibly in ink.**

**SECTION A**

**Read carefully**

1   Check that the answer sheet provided is for **Chemistry Advanced Higher (Section A)**.

2   For this section of the examination you must use an **HB pencil** and, where necessary, an eraser.

3   Check that the answer sheet you have been given has **your name**, **date of birth**, **SCN** (Scottish Candidate Number) and **Centre Name** printed on it.

    Do not change any of these details.

4   If any of this information is wrong, tell the Invigilator immediately.

5   If this information is correct, **print** your name and seat number in the boxes provided.

6   The answer to each question is **either** A, B, C or D.  Decide what your answer is, then, using your pencil, put a horizontal line in the space provided (see sample question below).

7   There is **only one correct** answer to each question.

8   Any rough working should be done on the question paper or the rough working sheet, **not** on your answer sheet.

9   At the end of the exam, put the **answer sheet for Section A inside the front cover of your answer book**.

**Sample Question**

To show that the ink in a ball-pen consists of a mixture of dyes, the method of separation would be

> A   chromatography
>
> B   fractional distillation
>
> C   fractional crystallisation
>
> D   filtration.

The correct answer is **A**—chromatography.  The answer **A** has been clearly marked in **pencil** with a horizontal line (see below).

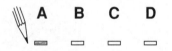

**Changing an answer**

If you decide to change your answer, carefully erase your first answer and using your pencil, fill in the answer you want.  The answer below has been changed to **D**.

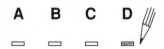

1. An atom has the electronic configuration

   $$1s^2\, 2s^2\, 2p^6\, 3s^2\, 3p^1$$

   What is the charge of the most likely ion formed from this atom?

   A   −1

   B   +1

   C   +2

   D   +3

2. The electronic configurations, **X** and **Y**, for two uncharged atoms of sodium are as follows.

   $X \quad 1s^2\, 2s^2\, 2p^6\, 3s^1$
   $Y \quad 1s^2\, 2s^2\, 2p^6\, 4s^1$

   Which of the following statements is true?

   A   **X** is an excited state.

   B   Both **X** and **Y** have vacant 2d orbitals.

   C   Energy is absorbed in changing **Y** to **X**.

   D   Less energy is required to ionise **Y** compared to **X**.

3. A Lewis base may be regarded as a substance which is capable of donating an unshared pair of electrons to form a covalent bond.

   Which of the following could act as a Lewis base?

   A   $Co^{3+}$

   B   $PH_3$

   C   $BCl_3$

   D   $NH_4^+$

4. Silicon can be converted into an n-type semiconductor by adding

   A   boron

   B   carbon

   C   arsenic

   D   aluminium.

5. Which of the following statements referring to the structures of sodium chloride and caesium chloride is correct?

   A   There are eight chloride ions surrounding each sodium ion.

   B   There are eight chloride ions surrounding each caesium ion.

   C   The chloride ions are arranged tetrahedrally round the sodium ions.

   D   The chloride ions are arranged tetrahedrally round the caesium ions.

6. The transition metal salts, $MnF_2$, $FeF_2$ and $CoF_2$, have identical crystal structures because the metal ions have

   A   similar radii

   B   similar colours

   C   the same nuclear charge

   D   the same number of d electrons.

7. Which of the following hydrides, when added to water, would give the most acidic solution?

   A   Sodium hydride

   B   Magnesium hydride

   C   Silicon hydride

   D   Sulphur hydride

8. Sodium hydride reacts with sodium sulphate as shown.

   $$4NaH\ +\ Na_2SO_4\ \rightarrow\ 4NaOH\ +\ Na_2S$$

   This reaction demonstrates sodium hydride acting as

   A   a base

   B   an acid

   C   a reducing agent

   D   an oxidising agent.

   **[Turn over**

9. Three elements, **X**, **Y** and **Z**, are in the same period of the Periodic Table.

   The oxide of **X** is amphoteric, the oxide of **Y** is basic and the oxide of **Z** is acidic.

   Which of the following shows the elements arranged in order of increasing atomic number?

   A  **Y, X, Z**

   B  **Y, Z, X**

   C  **Z, X, Y**

   D  **X, Y, Z**

10. Which of the following involves oxidation?

    A  $MnO_4^- \rightarrow MnO_4^{2-}$

    B  $Ag^+ \rightarrow [Ag(NH_3)_2]^+$

    C  $[Fe(CN)_6]^{4-} \rightarrow [Fe(CN)_6]^{3-}$

    D  $[Ni(H_2O)_6]^{2+} \rightarrow [Ni(CN)_4]^{2-}$

11. The number of unpaired electrons in a gaseous $Ni^{2+}$ ion is

    A  0

    B  2

    C  4

    D  6.

12.             $P + Q \rightleftharpoons R + S$

    At 298 K the equilibrium constant for this reaction is $1\cdot2 \times 10^{10}$.

    Which of the following is true?

    A  The value of $\Delta S^\circ$ must be positive.

    B  The value of $\Delta G^\circ$ must be positive.

    C  Adding a catalyst will change the equilibrium constant.

    D  Increasing the concentration of P will not change the equilibrium constant.

13.

    $$CH_3COOH + C_2H_5OH \rightleftharpoons CH_3COOC_2H_5 + H_2O$$

    The above reaction can be said to have reached equilibrium when

    A  the equilibrium constant K is equal to 1

    B  the reaction between the acid and the alcohol has stopped

    C  the concentrations of the products equal those of the reactants

    D  the rate of production of ethyl ethanoate equals its rate of hydrolysis.

14. When sulphur dioxide and oxygen react the following equilibrium is established.

    $$2SO_2(g) + O_2(g) \rightleftharpoons 2SO_3(g)$$

    The equilibrium constant for the reaction is 3300 at 630 °C and 21 at 850 °C.

    Which line in the table is correct for the reaction?

    |   | Sign of $\Delta H$ | Product yield as temperature increases |
    |---|---|---|
    | A | + | decreases |
    | B | + | increases |
    | C | – | decreases |
    | D | – | increases |

15. $500\,cm^3$ of $0\cdot022$ mol$\,l^{-1}$ hydrochloric acid is mixed with $500\,cm^3$ of $0\cdot020$ mol$\,l^{-1}$ sodium hydroxide solution. The pH of the resulting solution will be

    A  2

    B  3

    C  4

    D  5.

16. The Bronsted-Lowry definition of a base is a substance which acts as a

    A  proton donor to form a conjugate acid

    B  proton donor to form a conjugate base

    C  proton acceptor to form a conjugate acid

    D  proton acceptor to form a conjugate base.

17. The mean bond enthalpy of the N–H bond is equal to one third of the value of ΔH for which change of the following changes?

   A   $NH_3(g) \rightarrow N(g) + 3H(g)$

   B   $2NH_3(g) \rightarrow N_2(g) + 3H_2(g)$

   C   $NH_3(g) \rightarrow \frac{1}{2}N_2(g) + 1\frac{1}{2}H_2(g)$

   D   $2NH_3(g) + 1\frac{1}{2}O_2(g) \rightarrow N_2(g) + 3H_2O(g)$

18. The entropy of a perfect crystal is zero at

   A   $0\,K$

   B   $25\,K$

   C   $273\,K$

   D   $298\,K$.

19. Which of the following reactions results in a **decrease** in entropy?

   A   $N_2O_4(g) \rightarrow 2NO_2(g)$

   B   $N_2(g) + 3H_2(g) \rightarrow 2NH_3(g)$

   C   $CaCO_3(s) \rightarrow CaO(s) + CO_2(g)$

   D   $C(s) + H_2O(g) \rightarrow CO(g) + H_2(g)$

20. Which of the following is **not** a required condition for measuring standard electrode potentials?

   A   Volume of 1 litre

   B   Temperature of 298 K

   C   Concentration of $1\,mol\,l^{-1}$

   D   Pressure of 1 atmosphere

21.

metal X          Ag

salt bridge

$1\,mol\,l^{-1}\,X^{2+}(aq)$          $1\,mol\,l^{-1}\,Ag^+(aq)$

The $E^\circ$ values are

   $X^{2+}(aq) + 2e^- \rightarrow X(s)$          $E^\circ = -0{\cdot}23\,V$

   $Ag^+(aq) + e^- \rightarrow Ag(s)$          $E^\circ = 0{\cdot}80\,V$

In the above cell, which of the following is reduced?

   A   $X(s)$

   B   $Ag(s)$

   C   $X^{2+}(aq)$

   D   $Ag^+(aq)$

22. Under standard conditions, the emf of the cell

   $Al(s)\,|\,Al^{3+}(aq)\,|\,Cu^{2+}(aq)\,|\,Cu(s)$

   would be

   A   $1{\cdot}34\,V$

   B   $2{\cdot}02\,V$

   C   $2{\cdot}34\,V$

   D   $4{\cdot}38\,V$.

23. For a cell in which the following reaction occurs

   $X(s) + 2Y^+(aq) \rightarrow X^{2+}(aq) + 2Y(s)$

   the $E^\circ$ value is $1{\cdot}5\,V$.

   $\Delta G^\circ$ for the reaction, per mole of X, is

   A   $-289{\cdot}5\,kJ\,mol^{-1}$

   B   $-144{\cdot}8\,kJ\,mol^{-1}$

   C   $+144{\cdot}8\,kJ\,mol^{-1}$

   D   $+289{\cdot}5\,kJ\,mol^{-1}$.

**[Turn over**

**24.**

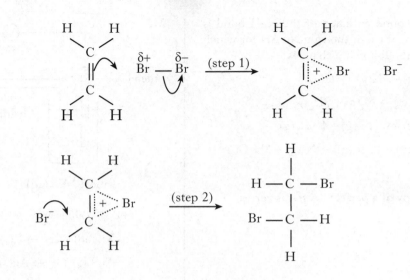

The two steps in the reaction mechanism shown can be described as

A   ethene acting as a nucleophile and $Br^-$ acting as a nucleophile

B   ethene acting as a nucleophile and $Br^-$ acting as an electrophile

C   ethene acting as an electrophile and $Br^-$ acting as a nucleophile

D   ethene acting as an electrophile and $Br^-$ acting as an electrophile.

---

**25.** In the homologous series of alkanols, increase in chain length from $CH_3OH$ to $C_{10}H_{21}OH$ is accompanied by

A   increased volatility and increased solubility in water

B   increased volatility and decreased solubility in water

C   decreased volatility and decreased solubility in water

D   decreased volatility and increased solubility in water.

**26.** Which of the following is **not** caused by hydrogen bonding?

A   The low density of ice compared to water

B   The solubility of methoxymethane in water

C   The higher boiling point of methanol compared to ethane

D   The higher melting point of hydrogen compared to helium

**27.** A compound $C_3H_8O$ does **not** react with sodium and is **not** reduced by lithium aluminium hydride. It is likely to be an

A   acid

B   ether

C   alcohol

D   aldehyde.

**28.** Which of the following is least acidic?

A   $CH_3OH$

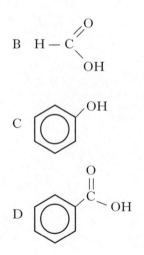

**29.** Which statement about ethanol and its isomeric ether is true?

They

A have similar volatilities

B have similar infra-red spectra

C form the same products when burned in excess oxygen

D form the same products when reacted with acidified dichromate.

**30.** When 2-bromobutane is reacted with potassium cyanide and the compound formed is hydrolysed with dilute acid, the final product is

A butanoic acid

B pentanoic acid

C 2-methylbutanoic acid

D 2-methylpropanoic acid.

**31.** Which of the following compounds would liberate one mole of hydrogen gas if one mole of it reacts with excess sodium?

A $C_2H_5OH$

B $CH_3CHO$

C $CH_3COOH$

D $HOCH_2CH_2OH$

**32.** Two isomeric esters, **X** and **Y**, have the molecular formula $C_4H_8O_2$. Ester **X** on hydrolysis with sodium hydroxide solution gives $CH_3CH_2COONa$, and ester **Y** on similar treatment gives $CH_3CH_2OH$.

Which line in the table shows the correct names of **X** and **Y**?

|   | X | Y |
|---|---|---|
| A | propyl methanoate | ethyl ethanoate |
| B | methyl propanoate | ethyl ethanoate |
| C | methyl propanoate | ethyl methanoate |
| D | propyl methanoate | methyl propanoate |

**33.** A white crystalline compound, soluble in water, was found to react with both dilute hydrochloric acid and sodium hydroxide solution.

Which of the following might it have been?

A $C_6H_5OH$

B $C_6H_5NH_2$

C $C_6H_5COOH$

D $H_2NCH_2COOH$

**34.** Which of the following amines has the lowest boiling point?

A $C_4H_9NH_2$

B $C_3H_7NHCH_3$

C $C_2H_5NHC_2H_5$

D $C_2H_5N(CH_3)_2$

**35.** Spectral studies of an organic compound indicated the presence of a di-substituted benzene ring, two methyl groups and a molecular weight of 134.

Which of the following is a possible structure for the compound?

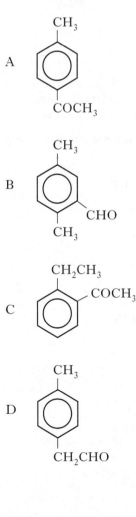

[**Turn over**

**36.** Which of the following molecules does **not** exhibit optical isomerism?

A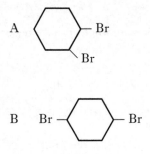

B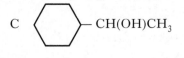

C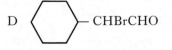

D

**37.** Which of the following could **not** exist in isomeric forms?

A  $C_2F_4$

B  $C_3H_6$

C  $C_3H_7Br$

D  $C_2H_4Cl_2$

**38.** Elemental analysis of an organic compound showed it contained 70·6% carbon, 23·5% oxygen and 5·9% hydrogen by mass.

The structural formula of the compound could be

A  $CH_2CHCH_2COOH$

B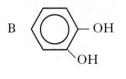

C

D

**39.** An organic compound with empirical formula, $C_2H_4O$, has major peaks at $1715\,cm^{-1}$ and $3300\,cm^{-1}$ in its infrared spectrum.

The structural formula of the compound could be

A  $CH_3CHO$

B  $CH_3COOH$

C  $CH_3COOCH_2CH_3$

D  $CH_3CH_2CH_2COOH$.

**40.** A drug containing a carboxyl group can bind to an amino group on a receptor site in three different ways.

| **Hydrogen-bond acceptor** | **Hydrogen-bond donor** | **Ionic interaction** |
|---|---|---|
| 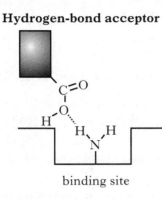 | 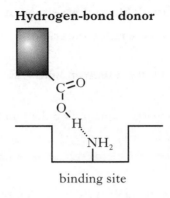 | 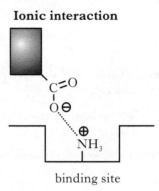 |
| binding site | binding site | binding site |

The drug with the following structure

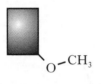

could bind to the same site

A   only by ionic interaction

B   only as a hydrogen-bond donor

C   only as a hydrogen-bond acceptor

D   both as a hydrogen-bond donor and acceptor.

*[END OF SECTION A]*

**Candidates are reminded that the answer sheet for Section A MUST be placed INSIDE the front cover of your answer book.**

**[Turn over for SECTION B on *Page ten***

**SECTION B**

**60 marks are available in this section of the paper.**

**All answers must be written clearly and legibly in ink.**

*Marks*

1. (a) What is the arrangement of the **electron pairs** around the iodine atom in an $IF_5$ molecule?

    1

   (b) By considering the electron pairs, explain why the bond angle in $BF_3$ is greater than the bond angle in $NF_3$.

    1

    (2)

2. An aqueous solution of the compound $[CoCl_2(NH_3)_4]Cl$ gave the following **transmittance** spectrum.

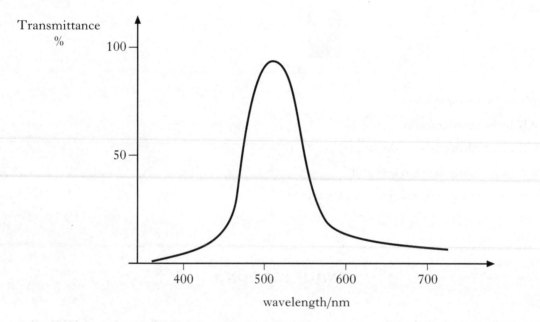

wavelength/nm

   (a) From the above spectrum, deduce the colour of the solution.

    1

   (b) The solution contains the complex ion $[CoCl_2(NH_3)_4]^+$.

    (i) What is the oxidation number of cobalt in this complex ion?

    1

    (ii) Name this complex ion.

    1

    (iii) Write down the electronic configuration of cobalt in this complex ion in terms of s, p and d orbitals.

    1

    (4)

3. Some metal salts emit light when heated in a Bunsen flame.

   Lithium nitrate changes the flame colour to crimson.

   Magnesium nitrate has no effect on the flame colour.

   (a) Explain, in terms of electrons, why some metal salts emit light when heated in a Bunsen flame.

    1

   (b) Suggest why magnesium nitrate has no effect on the flame colour.

    1

   (c) Calculate the energy, in $kJ\,mol^{-1}$, associated with crimson light of wavelength 671 nm.

    2

    (4)

*Marks*

**4.** *cis*-Platin is a highly successful anti-cancer drug. The formula for *cis*-platin is $[Pt(NH_3)_2Cl_2]$.

(a) *cis*-Platin works by forming a complex with parts of a DNA molecule. These parts of the DNA form bonds through nitrogen atoms to $Pt(NH_3)_2$ as shown below.

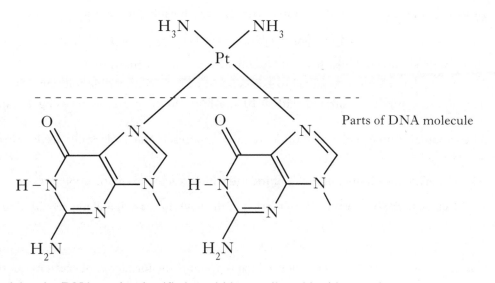

Parts of DNA molecule

    (i)   Explain why DNA can be classified as a bidentate ligand in this complex.      1

    (ii)  What feature of the DNA makes it suitable as a ligand?      1

(b) Draw a possible structure for the geometric isomer of *cis*-platin.      1

    **(3)**

**5.** The equation for the decomposition of ammonium dichromate is

$$(NH_4)_2Cr_2O_7(s) \longrightarrow N_2(g) + 4H_2O(\ell) + Cr_2O_3(s)$$

Consider the following data for the reaction at 298 K.

| Substance | $\Delta H_f^\circ/kJ\ mol^{-1}$ | $S^\circ/J\ K^{-1}mol^{-1}$ |
|---|---|---|
| $(NH_4)_2Cr_2O_7(s)$ | −1806 | 336 |
| $N_2(g)$ | 0 | 192 |
| $H_2O(\ell)$ | −286 | 70 |
| $Cr_2O_3(s)$ | −1140 | 81 |

(a) For the decomposition of $(NH_4)_2Cr_2O_7$, calculate

    (i)   $\Delta H^\circ$      1

    (ii)  $\Delta S^\circ$      1

    (iii) $\Delta G^\circ$.      2

(b) Chromium burns in excess oxygen to form chromium(III) oxide. From the information in the table, deduce a value for the enthalpy of combustion of chromium.      1

    **(5)**

**[Turn over**

*Marks*

6. Sodium hypochlorite, NaClO, is the active ingredient in household bleach. The concentration of the hypochlorite ion, ClO⁻, can be determined in two stages.

   In stage 1, an acidified iodide solution is added to a solution of the bleach and iodine is formed.

   $$ClO^-(aq) + 2I^-(aq) + 2H^+(aq) \rightarrow I_2(aq) + Cl^-(aq) + H_2O(\ell)$$

   In stage 2, the iodine formed is titrated with sodium thiosulphate solution.

   $$2S_2O_3^{2-}(aq) + I_2(aq) \rightarrow 2I^-(aq) + S_4O_6^{2-}(aq)$$

   $10.0 \, cm^3$ of a household bleach was diluted to $250 \, cm^3$ in a standard flask.

   $25.0 \, cm^3$ of this solution was added to excess acidified potassium iodide solution.

   The solution was then titrated with $0.10 \, mol \, l^{-1}$ sodium thiosulphate using an appropriate indicator.

   The volume of thiosulphate solution required to reach the end point of the titration was $20.5 \, cm^3$.

   (a) Calculate the number of moles of **iodine** which reacted in the titration.    1

   (b) Calculate the concentration, in $mol \, l^{-1}$, of the ClO⁻ in the original household bleach.    2

   (3)

7. Consider the Born-Haber cycle below which represents the formation of caesium fluoride.

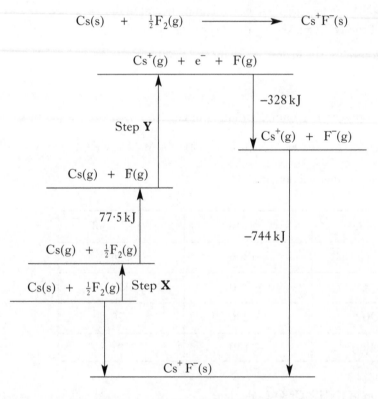

   (a) Use the Data Booklet to find the enthalpy values for Step **X** and Step **Y**.    2

   (b) Name the enthalpy change that has the value −744 kJ in this cycle.    1

   (c) Use this Born-Haber cycle to calculate the enthalpy of formation of caesium fluoride in $kJ \, mol^{-1}$.    1

   (4)

*Marks*

8. In a PPA, a student added $50\,cm^3$ of an aqueous iodine solution to $50\,cm^3$ of cyclohexane in a separating funnel. After shaking thoroughly, the funnel was left until the following equilibrium was established.

$$I_2(aqueous) \rightleftharpoons I_2(cyclohexane)$$

Two layers were formed, each containing dissolved iodine. $10.0\,cm^3$ of each layer was titrated with sodium thiosulphate solution until the end point was reached.

The cyclohexane layer required $18.8\,cm^3$ of $0.025\,mol\,l^{-1}$ sodium thiosulphate.

The aqueous layer required $10.5\,cm^3$ of $0.050\,mol\,l^{-1}$ sodium thiosulphate.

(a) Which indicator is used to show that the end point has been reached?      1

(b) Calculate the concentration of iodine in

    (i) the cyclohexane layer      1

   (ii) the aqueous layer.      1

(c) Calculate the partition coefficient for iodine between the two solvents.      1

(d) If $100\,cm^3$ of cyclohexane had been used instead of the $50\,cm^3$, what effect would this have on

    (i) the concentration of iodine in the aqueous layer      1

   (ii) the value of the partition coefficient?      1

**(6)**

**[Turn over**

*Marks*

9.  Hydrofluoric acid, HF, is a weak acid.

$$HF(aq) \; + \; H_2O(\ell) \; \rightleftharpoons \; H_3O^+(aq) \; + \; F^-(aq)$$

A student neutralised $25\,cm^3$ of hydrofluoric acid solution with sodium hydroxide solution and followed the reaction by measuring the pH.

The graph obtained for this reaction is shown below.

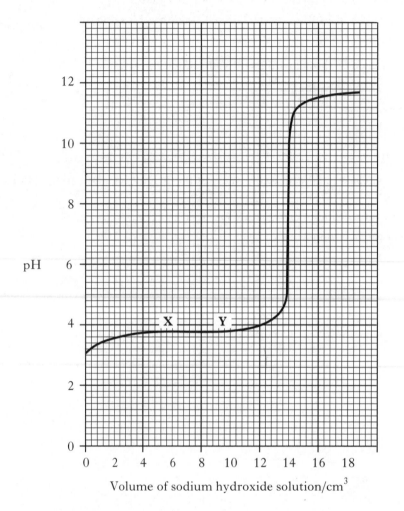

*pH*

Volume of sodium hydroxide solution/cm³

(a)  Write the expression for the dissociation constant, $K_a$, of hydrofluoric acid.    **1**

(b)  When exactly half the acid has been neutralised, $pK_a = pH$.

**Using only information from the graph**, deduce $pK_a$ and thus calculate $K_a$ for hydrofluoric acid.    **2**

(c)  The region **XY** on the graph is sometimes referred to as the buffer region.

Apart from HF, what else is present in the solution which enables it to act as a buffer?    **1**

(d)

| Indicator | $pK_{In}$ |
|---|---|
| Methyl orange | 3·7 |
| Alizarin red | 6·6 |
| Cresol red | 8·0 |
| Alizarin yellow | 11·1 |

Which of the above indicators could be used to detect the end point of this neutralisation reaction?    **1**

**(5)**

*Marks*

10. A mixture of butan-1-ol and butan-2-ol can be synthesised from 1-bromobutane in a two stage process.

$$CH_3CH_2CH_2CH_2Br \xrightarrow[\textbf{Stage 1}]{KOH/C_2H_5OH} CH_3CH_2CH=CH_2 \xrightarrow[\textbf{Stage 2}]{H_2O/H^+} \begin{array}{c} \text{butan-1-ol} \\ + \\ \text{butan-2-ol} \end{array}$$

(a) What type of reaction is taking place in **Stage 1**?                                                                1

(b) The bonding in but-1-ene can be described in terms of $sp^2$ and $sp^3$ hybridisation and sigma and pi bonds.

  (i) What is meant by $sp^2$ hybridisation?                                                                1

  (ii) What is the difference in the way atomic orbitals overlap to form sigma and pi bonds?                                                                1

(c) Draw a structural formula for the major product of **Stage 2**.                                                                1

(d) 1-Bromobutane reacts with hydroxide ions in a nucleophilic substitution reaction to produce butan-1-ol. The following results were obtained for this reaction.

| Experiment | [1-Bromobutane]/$mol\,l^{-1}$ | [$OH^-$]/$mol\,l^{-1}$ | Initial rate/$mol\,l^{-1}\,s^{-1}$ |
|:---:|:---:|:---:|:---:|
| 1 | 0·25 | 0·10 | $3·3 \times 10^{-6}$ |
| 2 | 0·50 | 0·10 | $6·6 \times 10^{-6}$ |
| 3 | 0·50 | 0·20 | $1·3 \times 10^{-5}$ |

  (i) What is the overall order of this reaction?                                                                1

  (ii) Calculate a value for the rate constant of this reaction, giving the appropriate units.                                                                2

  (iii) Outline the mechanism for this nucleophilic substitution reaction using structural formulae.                                                                2

                                                                                                                **(9)**

11. A student devised the following reaction sequence starting from propan-1-ol, $C_3H_7OH$.

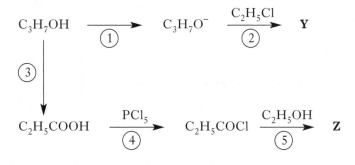

(a) Name a suitable reagent to carry out

  (i) Step ①

  (ii) Step ③.                                                                2

(b) Name **Y**.                                                                1

(c) Draw a structural formula for **Z**.                                                                1

                                                                                                                **(4)**

**[Turn over**

*Marks*

12. In a PPA, propanone reacts with 2,4-dinitrophenylhydrazine to make the 2,4-dinitrophenylhydrazone derivative as shown below.

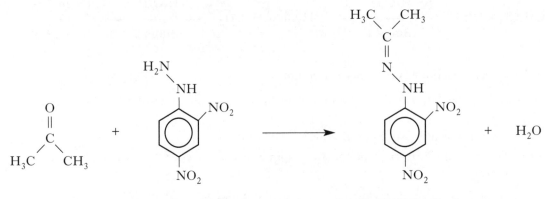

propanone    2,4-dinitrophenylhydrazine    2,4-dinitrophenylhydrazone derivative

(a) What type of reaction is this?      1

(b) The 2,4-dinitrophenylhydrazone derivative formed in the reaction is impure.

   (i) How would the derivative be purified?      1

   (ii) How can the technique of derivative formation be used to identify an unknown ketone?      1

(c) Propanone has an isomer. The shortened structural formula of this isomer is $CH_3CH_2CHO$.

   (i) Which chemical reagent could be used to distinguish between propanone and this isomer and what would be the result?      1

   (ii) Nuclear magnetic resonance spectroscopy can also be used to distinguish between these two isomers. The proton nmr spectrum for $CH_3CH_2CHO$ is shown.

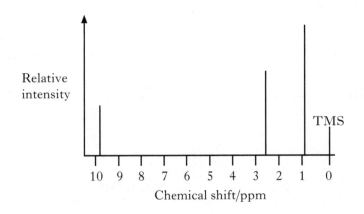

Sketch the proton nmr spectrum you would obtain for propanone.      1

*Marks*

**12.** (*c*)  **(continued)**

    (iii)  A simplified mass spectrum for propanone is shown below.

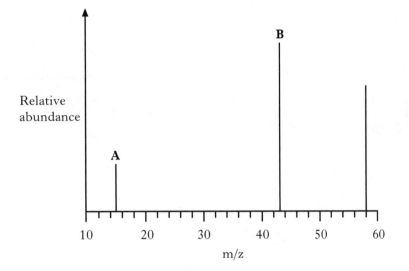

Identify the ion fragments responsible for peaks **A** and **B**.     **1**

    **(6)**

**[Turn over for Question 13 on *Page eighteen***

*Marks*

**13.** A student devised the following reaction scheme starting with benzene.

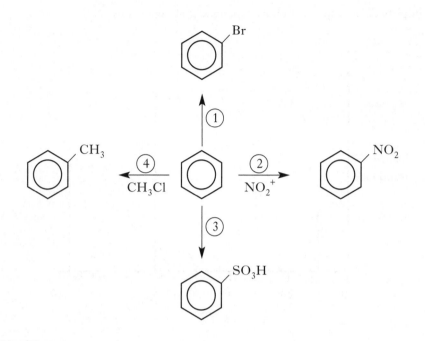

(*a*)  What type of reaction does benzene undergo in reactions ①–④ ?                                    1

(*b*)  Name a suitable reagent and catalyst for reaction ① .                                             1

(*c*)  Reaction ② involves nitration of benzene.
       Which reagents are used to produce the $NO_2^+$ ion?                                              1

(*d*)  What is the molecular formula for the product of reaction ③ ?                                     1

(*e*)  The product of reaction ④ was reacted with bromine in the presence of light.
       Draw a structural formula for an organic product of this reaction.                               1

                                                                                                     **(5)**

*[END OF QUESTION PAPER]*

# 2009

[BLANK PAGE]

# X012/701

NATIONAL
QUALIFICATIONS
2009

WEDNESDAY, 3 JUNE
9.00 AM – 11.30 AM

CHEMISTRY
ADVANCED HIGHER

Reference may be made to the Chemistry Higher and Advanced Higher Data Booklet .

**SECTION A — 40 marks**

Instructions for completion of **SECTION A** are given on page two.

For this section of the examination you must use an **HB pencil**.

**SECTION B — 60 marks**

All questions should be attempted.

**Answers must be written clearly and legibly in ink.**

## SECTION A

**Read carefully**

1 Check that the answer sheet provided is for **Chemistry Advanced Higher (Section A)**.

2 For this section of the examination you must use an **HB pencil** and, where necessary, an eraser.

3 Check that the answer sheet you have been given has **your name**, **date of birth**, **SCN** (Scottish Candidate Number) and **Centre Name** printed on it.

    Do not change any of these details.

4 If any of this information is wrong, tell the Invigilator immediately.

5 If this information is correct, **print** your name and seat number in the boxes provided.

6 The answer to each question is **either** A, B, C or D. Decide what your answer is, then, using your pencil, put a horizontal line in the space provided (see sample question below).

7 There is **only one correct** answer to each question.

8 Any rough working should be done on the question paper or the rough working sheet, **not** on your answer sheet.

9 At the end of the exam, put the **answer sheet for Section A inside the front cover of your answer book**.

**Sample Question**

To show that the ink in a ball-pen consists of a mixture of dyes, the method of separation would be

    A  chromatography

    B  fractional distillation

    C  fractional crystallisation

    D  filtration.

The correct answer is **A**—chromatography. The answer **A** has been clearly marked in **pencil** with a horizontal line (see below).

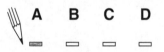

**Changing an answer**

If you decide to change your answer, carefully erase your first answer and using your pencil, fill in the answer you want. The answer below has been changed to **D**.

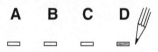

1. The diagram shows one of the series of lines in the hydrogen emission spectrum.

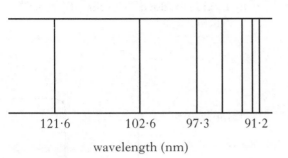

wavelength (nm)

Each line

A represents an energy level within a hydrogen atom

B results from an electron moving to a higher energy level

C lies within the visible part of the electromagnetic spectrum

D results from an excited electron dropping to a lower energy level.

2. Which of the following compounds shows most covalent character?

A $CH_4$

B NaH

C $NH_3$

D $PH_3$

3. In which of the following species is a dative covalent bond present?

A $H_3O^+$

B $H_2O$

C $OH^-$

D $O_2$

4. Which of the following diagrams best represents the arrangement of electron pairs around the central iodine atom in the $I_3^-$ ion?

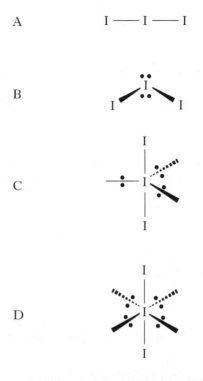

5. When a voltage is applied to an n-type semiconductor, which of the following migrate through the lattice?

A Electrons

B Negative ions

C Positive holes

D Both electrons and positive holes

6. Which of the following compounds would produce fumes of hydrogen chloride when added to water?

A LiCl

B $MgCl_2$

C $PCl_3$

D $CCl_4$

**[Turn over**

7. Zinc oxide reacts as shown.

$ZnO(s) + 2HCl(aq) \rightarrow ZnCl_2(aq) + H_2O(\ell)$

$ZnO(s) + 2NaOH(aq) + H_2O(\ell) \rightarrow Na_2Zn(OH)_4(aq)$

This shows that zinc oxide is

A  basic

B  acidic

C  neutral

D  amphoteric.

8. The correct formula for the tetraamminedichlorocopper(II) complex is

A  $[Cu(NH_3)_4Cl_2]^{2-}$

B  $[Cu(NH_3)_4Cl_2]$

C  $[Cu(NH_3)_4Cl_2]^{2+}$

D  $[Cu(NH_3)_4Cl_2]^{4+}$.

9. Which of the following aqueous solutions contains the **greatest** number of **negatively** charged ions?

A  $500\,cm^3$ $0.10\,mol\,l^{-1}$ $Na_2SO_4(aq)$

B  $250\,cm^3$ $0.12\,mol\,l^{-1}$ $BaCl_2(aq)$

C  $300\,cm^3$ $0.15\,mol\,l^{-1}$ $KI(aq)$

D  $400\,cm^3$ $0.10\,mol\,l^{-1}$ $Zn(NO_3)_2(aq)$

10. When one mole of phosphorus pentachloride was heated to 523 K in a closed vessel, 50% dissociated as shown.

$PCl_5(g) \rightleftharpoons PCl_3(g) + Cl_2(g)$

How many moles of gas were present in the equilibrium mixture?

A  0·5

B  1·0

C  1·5

D  2·0

11. Which of the following graphs shows the temperature change as $2\,mol\,l^{-1}$ sodium hydroxide is added to $25\,cm^3$ of $2\,mol\,l^{-1}$ hydrochloric acid?

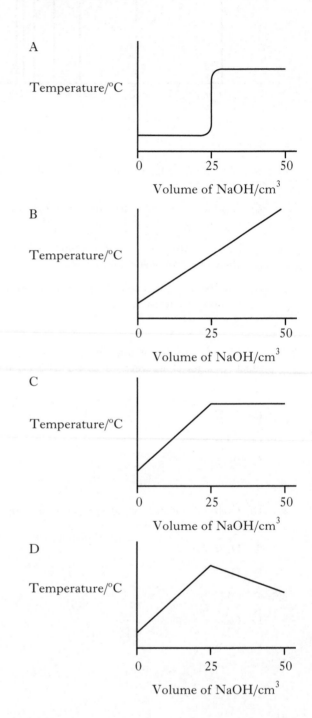

**12.** In the equilibrium $N_2O_4(g) \rightleftharpoons 2NO_2(g)$ the forward reaction is endothermic.

Which one of the following causes an increase in the value of the equilibrium constant?

A   The removal of $NO_2$

B   An increase of pressure

C   A decrease of temperature

D   An increase of temperature

**13.** In which of the following separation techniques is partition between two separate phases **not** a part of the process?

A   Recrystallisation of benzoic acid from hot water

B   Separation of alkanes using gas-liquid chromatography

C   Separation of plant dyes using paper chromatography

D   Solvent extraction of caffeine from an aqueous solution using dichloromethane

**14.** An aqueous solution of an organic acid, X, was shaken with chloroform until the following equilibrium was established.

$$X \text{ (water)} \rightleftharpoons X \text{ (chloroform)}$$

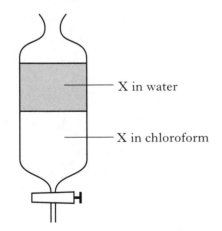

X in water

X in chloroform

$25.0 \, \text{cm}^3$ of the upper layer needed $20.0 \, \text{cm}^3$ of $0.050 \, \text{mol} \, l^{-1}$ NaOH(aq) for neutralisation.
$25.0 \, \text{cm}^3$ of the lower layer needed $13.3 \, \text{cm}^3$ of $0.050 \, \text{mol} \, l^{-1}$ NaOH(aq) for neutralisation.

The value of the partition coefficient is

A   0.67

B   1.25

C   1.50

D   1.88.

**15.** Which of the following would **not** be suitable to act as a buffer solution?

A   Boric acid and sodium borate

B   Nitric acid and sodium nitrate

C   Benzoic acid and sodium benzoate

D   Propanoic acid and sodium propanoate

**16.** Which of the following $0.01 \, \text{mol} \, l^{-1}$ aqueous solutions has the highest pH value?

A   Sodium fluoride

B   Sodium benzoate

C   Sodium propanoate

D   Sodium methanoate

**17.** Which of the following graphs shows the variation in $\Delta G°$ with temperature for a reaction which is always feasible?

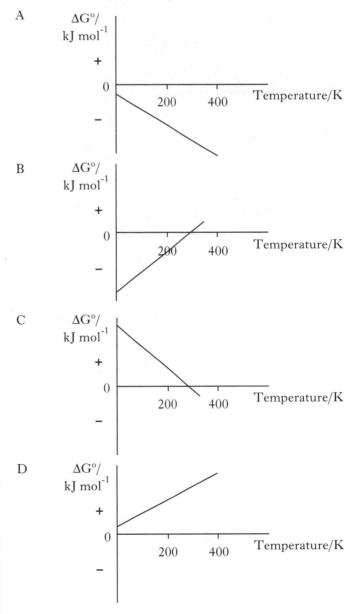

18. When water evaporates from a puddle which of the following applies?

    A   $\Delta H$ positive and $\Delta S$ positive

    B   $\Delta H$ positive and $\Delta S$ negative

    C   $\Delta H$ negative and $\Delta S$ positive

    D   $\Delta H$ negative and $\Delta S$ negative

19. For which of the following reactions would the value of $\Delta G° - \Delta H°$ be approximately zero?

    A   $CaCO_3(s) \rightarrow CaO(s) + CO_2(g)$

    B   $C(s) + H_2O(g) \rightarrow CO(g) + H_2(g)$

    C   $Zn(s) + 2H^+(aq) \rightarrow Zn^{2+}(aq) + H_2(g)$

    D   $Cu^{2+}(aq) + Mg(s) \rightarrow Mg^{2+}(aq) + Cu(s)$

20. For the reaction

    $$2NO(g) + Cl_2(g) \rightarrow 2NOCl(g)$$

    the rate equation is

    $$rate = k[NO][Cl_2].$$

    The overall order of this reaction is

    A   1

    B   2

    C   3

    D   5.

21. The following data refer to initial reaction rates obtained for the reaction

    $$X + Y + Z \rightarrow \text{products}$$

| Run | Relative concentrations | | | Relative initial rate |
|---|---|---|---|---|
|  | [X] | [Y] | [Z] |  |
| 1 | 1·0 | 1·0 | 1·0 | 0·3 |
| 2 | 1·0 | 2·0 | 1·0 | 0·6 |
| 3 | 2·0 | 2·0 | 1·0 | 1·2 |
| 4 | 2·0 | 1·0 | 2·0 | 0·6 |

    These data fit the rate equation

    A   Rate = $k[X]$

    B   Rate = $k[X][Y]$

    C   Rate = $k[X][Y]^2$

    D   Rate = $k[X][Y][Z]$

22. Which of the following is a propagation step in the chlorination of methane?

    A   $Cl_2 \rightarrow Cl^\bullet + Cl^\bullet$

    B   $CH_3^\bullet + Cl^\bullet \rightarrow CH_3Cl$

    C   $CH_3^\bullet + Cl_2 \rightarrow CH_3Cl + Cl^\bullet$

    D   $CH_4 + Cl^\bullet \rightarrow CH_3Cl + H^\bullet$

23. The hydrolysis of the halogenoalkane $(CH_3)_3CBr$ was found to take place by an $S_N1$ mechanism.

    The rate-determining step involved the formation of

    A

    B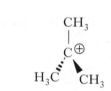

    C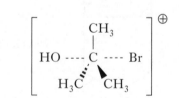

    D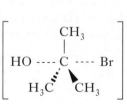

24. $OH^- + CO_2 \rightarrow HCO_3^-$

$C_2H_4 + Br_2 \rightarrow C_2H_4Br^+ + Br^-$

Which substances act as electrophiles in the above reactions?

A $OH^-$ and $Br_2$

B $OH^-$ and $C_2H_4$

C $CO_2$ and $Br_2$

D $CO_2$ and $C_2H_4$

25.

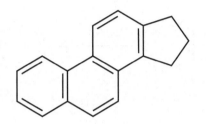

What is the molecular formula for the above structure?

A $C_{17}H_{11}$

B $C_{17}H_{14}$

C $C_{17}H_{17}$

D $C_{17}H_{20}$

26. Which line in the table is correct for the following hydrocarbon?

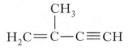

| | Number of σ bonds | Number of π bonds |
|---|---|---|
| A | 4 | 3 |
| B | 8 | 5 |
| C | 10 | 2 |
| D | 10 | 3 |

27. When but-2-ene is shaken with an aqueous solution of chlorine in potassium iodide, the structural formula(e) of the product(s) is/are

A
$$CH_3-\underset{\underset{I}{|}}{CH}-\underset{\underset{I}{|}}{CH}-CH_3$$

B
$$CH_3-\underset{\underset{Cl}{|}}{CH}-\underset{\underset{Cl}{|}}{CH}-CH_3$$

C
$$CH_3-\underset{\underset{Cl}{|}}{CH}-\underset{\underset{I}{|}}{CH}-CH_3 \text{ and } CH_2-\underset{\underset{I}{|}}{CH}-\underset{\underset{Cl}{|}}{CH_2}-CH_3$$

D
$$CH_3-\underset{\underset{Cl}{|}}{CH}-\underset{\underset{I}{|}}{CH}-CH_3 \text{ and } CH_3-\underset{\underset{Cl}{|}}{CH}-\underset{\underset{Cl}{|}}{CH}-CH_3$$

28. Which of the following reacts with ethanol to form the ethoxide ion?

A $Na(s)$

B $Na_2O(s)$

C $NaCl(aq)$

D $NaOH(aq)$

29. Which of the following is **not** a correct statement about ethoxyethane?

A It burns readily in air.

B It is isomeric with butan-2-ol.

C It has a higher boiling point than butan-2-ol.

D It is a very good solvent for many organic compounds.

[**Turn over**

30. Which of the following esters gives a secondary alcohol when hydrolysed?

A

$(CH_3)_3C-O-\overset{\overset{\displaystyle O}{\|}}{C}-H$

B

$CH_3-O-\overset{\overset{\displaystyle O}{\|}}{C}-CH(CH_3)_2$

C

$(CH_3)_2CH-O-\overset{\overset{\displaystyle O}{\|}}{C}-CH_3$

D

$(CH_3)_2CHCH_2-O-\overset{\overset{\displaystyle O}{\|}}{C}-CH_3$

31. Which of the following compounds could **not** be oxidised by acidified potassium dichromate solution?

A $CH_3CH_2CHO$

B $CH_3CH_2COOH$

C $CH_3CH_2CH_2OH$

D $CH_3CH(OH)CH_3$

32. Which of the following will react with dilute sodium hydroxide solution?

A $CH_3CHOHCH_3$

B $CH_3CH=CH_2$

C $CH_3COOCH_3$

D $CH_3CH_2OCH_3$

33. Which of the following molecules is planar?

A Hexane

B Cyclohexane

C Chlorobenzene

D Methylbenzene (toluene)

34. Which of the following compounds is soluble in water and reacts with both dilute hydrochloric acid and sodium hydroxide solution?

A $C_2H_5NH_2$

B $C_6H_5NH_2$

C $C_2H_5NH_3Cl$

D $HOOCCH_2NH_2$

35. Which of the following reactions is least likely to take place?

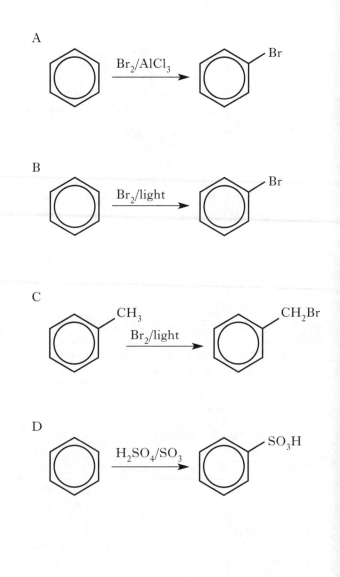

A
Br$_2$/AlCl$_3$ → Br

B
Br$_2$/light → Br

C
CH$_3$  Br$_2$/light → CH$_2$Br

D
H$_2$SO$_4$/SO$_3$ → SO$_3$H

**36.** In which of the following pairs does an aqueous solution of the first compound have a higher pH than an aqueous solution of the second?

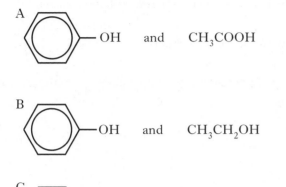

A   〇—OH   and   $CH_3COOH$

B   〇—OH   and   $CH_3CH_2OH$

C   〇—COOH   and   $HOCH_2CH_2OH$

D   〇—COOH   and   $CH_3OH$

**37.** Which of the following bases is the strongest?

  A   $C_2H_5NH_2$

  B   $(C_2H_5)_2NH$

  C   $C_6H_5NH_2$

  D   $(C_6H_5)_2NH$

**38.** Which line in the table shows a pair of optical isomers?

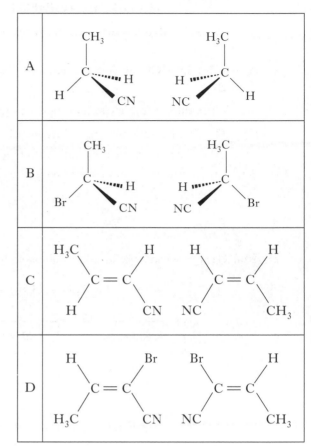

**39.**

Which atom in the above structure would be located **most** readily using X-ray crystallography?

  A   Carbon

  B   Hydrogen

  C   Iodine

  D   Oxygen

**40.** Antihistamines act by inhibiting the action of the inflammatory agent histamine in the body.

Antihistamines can be described as

  A   agonists

  B   receptors

  C   antagonists

  D   pharmacophores.

*[END OF SECTION A]*

**Candidates are reminded that the answer sheet for Section A MUST be placed INSIDE the front cover of your answer book.**

**SECTION B**

**60 marks are available in this section of the paper.**

**All answers must be written clearly and legibly in ink.**

*Marks*

1. A detector in a Geiger counter contains argon which ionises when nuclear radiation passes through it.

   (*a*)  Write the electronic configuration for argon in terms of s and p orbitals.    **1**

   (*b*)  The first ionisation energy of argon is $1530\,kJ\,mol^{-1}$.

       (i)   Calculate the wavelength of the radiation, in nm, corresponding to this energy.    **3**

       (ii)  Write the equation for the first ionisation of argon.    **1**

       **(5)**

2. Iron(III) oxide can be reduced to iron using hydrogen.

$$Fe_2O_3(s) + 3H_2(g) \rightarrow 2Fe(s) + 3H_2O(g)$$

| Substance | $\Delta H_f^\circ/kJ\,mol^{-1}$ | $S^\circ/J\,K^{-1}\,mol^{-1}$ |
|---|---|---|
| $Fe_2O_3(s)$ | −822 | 90 |
| $H_2(g)$ | 0 | 131 |
| $Fe(s)$ | 0 | 27 |
| $H_2O(g)$ | −242 | 189 |

For the reduction of iron(III) oxide with hydrogen, use the data in the table to calculate

   (*a*)  the standard entropy change, $\Delta S^\circ$    **1**

   (*b*)  the standard enthalpy change, $\Delta H^\circ$    **1**

   (*c*)  the theoretical temperature above which the reaction becomes feasible.    **2**

       **(4)**

**3.** The diagram, which is not drawn to scale, represents the processes involved in a thermochemical cycle.

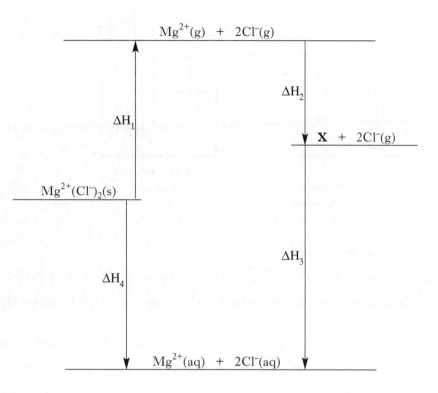

(a) What should be written in place of **X** to complete the diagram?    1

(b) What name is given to the enthalpy change represented by $\Delta H_1$?    1

(c) Calculate $\Delta H_3$ using information from the Data Booklet.    1

(d) Calculate $\Delta H_4$ using information from the Data Booklet.    1

     **(4)**

**4.** (a) Using the mean bond enthalpy values given in the Data Booklet, calculate the enthalpy change, in $kJ\,mol^{-1}$, for the reaction

$$H_2(g) + \tfrac{1}{2}O_2(g) \rightarrow H_2O(g)$$    3

(b) The value given in the Data Booklet for the standard enthalpy of combustion of hydrogen is different to that calculated in part (a).

Give the main reason for this difference.    1

     **(4)**

**[Turn over**

*Marks*

5.

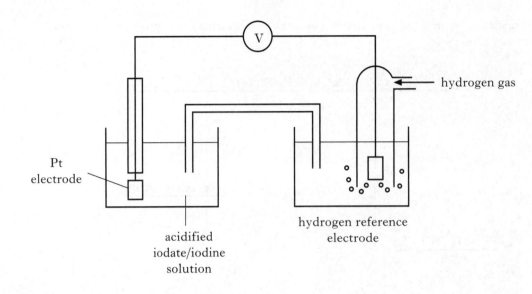

The above cell was set up under standard conditions.

(a)  What are the three standard conditions required for the hydrogen reference electrode?    1

(b)  Write an ion-electron equation for the reduction of iodate ions ($IO_3^-$) to iodine ($I_2$) in acidic conditions.    1

(c)  If the $E°$ value for the reduction of $IO_3^-$ to $I_2$ is $1·19\,V$, calculate the free energy change $\Delta G°$, in kJ per mole of $IO_3^-$, for the cell reaction.    3

**(5)**

6.  When an ant bites, it injects methanoic acid (HCOOH).

(a)  Methanoic acid is a weak acid.

$$HCOOH(aq) + H_2O(\ell) \rightleftharpoons HCOO^-(aq) + H_3O^+(aq)$$

(i)  What is the conjugate base of methanoic acid?    1

(ii)  Write the expression for the dissociation constant, $K_a$, of methanoic acid.    1

(b)  (i)  In a typical bite, an ant injects $3·6 \times 10^{-3}\,g$ of methanoic acid. Assuming that the methanoic acid dissolves in $1·0\,cm^3$ of water in the body, calculate the concentration of the methanoic acid solution in $mol\,l^{-1}$.    2

(ii)  Calculate the pH of this methanoic acid solution.    2

**(6)**

*Marks*

**7.** Iodine reacts with propanone as follows.

$$I_2 + CH_3COCH_3 \longrightarrow CH_3COCH_2I + HI$$

A possible mechanism for this reaction is

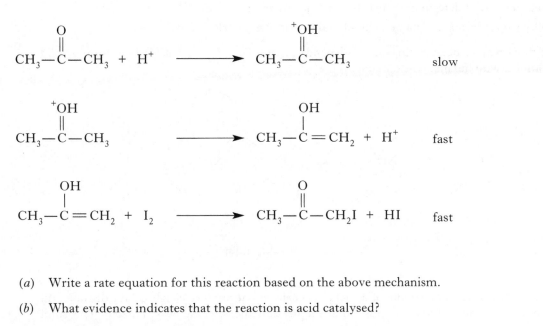

(a) Write a rate equation for this reaction based on the above mechanism.    1

(b) What evidence indicates that the reaction is acid catalysed?    1

(c) In a PPA the reaction was followed by withdrawing samples at regular intervals and adding them to sodium hydrogencarbonate solution.

   The concentration of iodine in these samples was then determined by titrating with a standard solution of sodium thiosulphate.

   (i) Why were the samples added to the sodium hydrogencarbonate solution?    1

   (ii) What indicator is used in the titration and what is the colour change at the end-point of the titration?    1

   **(4)**

**[Turn over**

*Marks*

8. Nickel can be determined quantitatively in a number of ways.

   (a) The method used in a PPA is volumetric analysis in which a buffered solution of nickel(II) ions is titrated against a standard solution of a complexing agent.

   Which complexing agent is used? **1**

   (b) Another way of determining nickel is by colorimetric analysis.

   Why would this be a suitable method of determining nickel(II) ions? **1**

   (c) A third way of determining nickel depends on the fact that nickel(II) ions form a solid complex with butanedione dioxime.

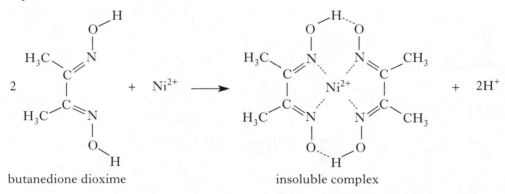

butanedione dioxime                              insoluble complex

Using this method, a sample of a nickel(II) salt was accurately weighed and dissolved in water. To this solution, excess butanedione dioxime solution was added. The solid complex formed was filtered, washed and then heated in an oven to constant mass.

   (i) Butanedione dioxime can act as a ligand.

   What property of butanedione dioxime allows it to act as a ligand? **1**

   (ii) What is the coordination number of the nickel(II) ion in the insoluble complex? **1**

   (iii) Which type of quantitative analysis has been carried out using this method? **1**

   (iv) During the process of heating to constant mass, the solid complex is cooled in a desiccator.

   Why is a desiccator used? **1**

   **(6)**

*Marks*

9. Compound **W** reacts in two steps to form compound **Y**.

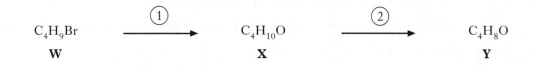

**Y** reacts with 2,4-dinitrophenylhydrazine solution (Brady's reagent) to form a yellow precipitate **Z**.

**Y** does not react with Fehling's solution, nor with Tollens' reagent.

(a) Identify compound **Y**.    1

(b) What type of reaction is occurring in step ①?    1

(c) What property of the yellow precipitate **Z** is measured and how is this used to confirm the identity of **Y**?    1

(d) Dehydration of compound **X** produces three unsaturated isomers of molecular formula $C_4H_8$. Two of these are **geometric** isomers.

Draw the structures of both **geometric** isomers and name each one.    2

**(5)**

10. N-Phenylethanamide can be prepared from benzene in three steps.

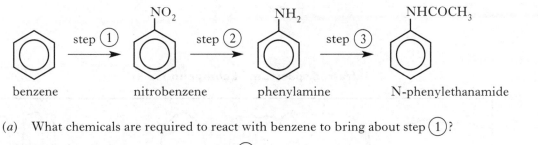

(a) What chemicals are required to react with benzene to bring about step ①?    1

(b) What type of reaction occurs in step ②?    1

(c) Suggest a reagent which could be used to bring about step ③.    1

**(3)**

**[Turn over**

11. Spectra of an organic compound **A** are shown below.

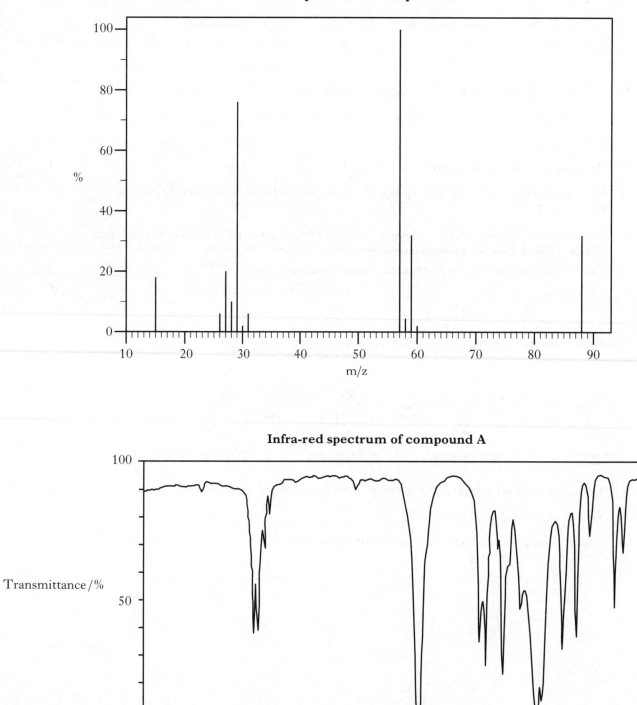

**Mass spectrum of compound A**

**Infra-red spectrum of compound A**

*Marks*

**11. (continued)**

**Proton nmr spectrum of compound A**

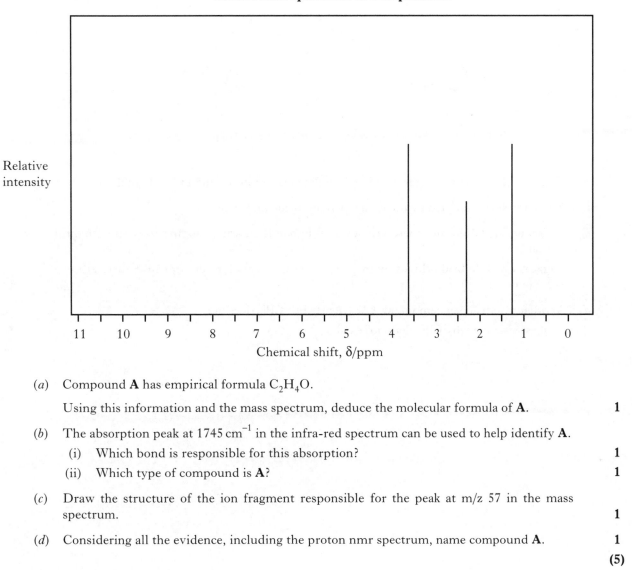

(a) Compound **A** has empirical formula $C_2H_4O$.

Using this information and the mass spectrum, deduce the molecular formula of **A**.   1

(b) The absorption peak at $1745 \, cm^{-1}$ in the infra-red spectrum can be used to help identify **A**.

   (i)   Which bond is responsible for this absorption?   1

   (ii)  Which type of compound is **A**?   1

(c) Draw the structure of the ion fragment responsible for the peak at m/z 57 in the mass spectrum.   1

(d) Considering all the evidence, including the proton nmr spectrum, name compound **A**.   1

             **(5)**

**[Turn over**

*Marks*

12. Many interhalogen compounds exist.  Two of these are iodine pentafluoride and iodine heptafluoride.

IF$_5$                           IF$_7$

(a) What are the oxidation states of iodine in iodine pentafluoride and iodine heptafluoride?    1

(b) Name the shape adopted by the iodine pentafluoride molecule.    1

(c) In iodine heptafluoride, there are seven I–F bonds in which iodine uses sp$^3$d$^3$ hybrid orbitals.

Suggest which hybrid orbitals iodine uses in iodine pentafluoride, in which there are five I–F bonds.    1

(d) Another interhalogen compound, ClF$_5$, exists but ClF$_7$ does not.

Suggest a reason why ClF$_7$ does not exist.    1

**(4)**

*Marks*

13. A superconductor, **X**, with a critical temperature of 95 K, was prepared by heating yttrium oxide, barium carbonate and copper oxide at high temperatures.

   (*a*) Copy the axes shown and sketch a graph to show how the **electrical resistance** of **X** varies with temperature.

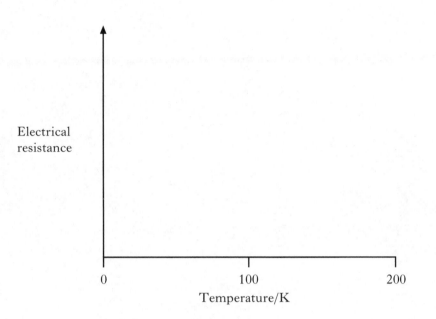

1

   (*b*) (i) **X** contains 13·4% yttrium, 41·2% barium, 28·6% copper and 16·8% oxygen.

   Assuming that the relative atomic mass of yttrium is 88·9, show by calculation that the empirical formula for **X** is $YBa_2Cu_3O_7$. 2

   (ii) Assuming that the oxidation states of yttrium, barium and oxygen are +3, +2 and −2 respectively, calculate the **average** oxidation state of copper in **X**. 1

   (iii) When all the copper(III) initially present in **X** is reduced to copper(II), compound **Z** is produced. The oxidation states of the other three elements do not change nor does the mole ratio of the **metals**.

   Suggest an empirical formula for **Z**. 1

   **(5)**

[*END OF QUESTION PAPER*]

[BLANK PAGE]

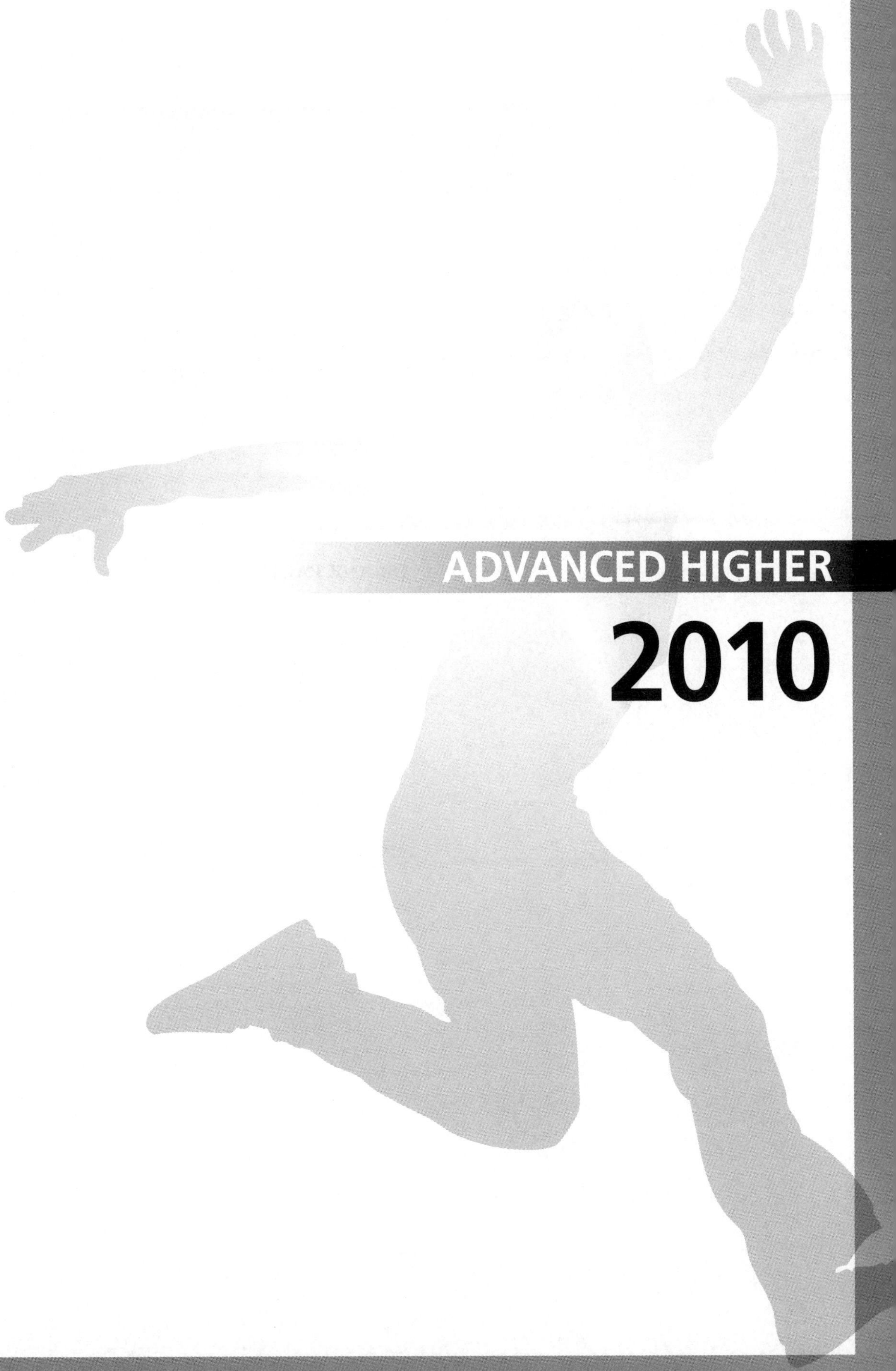

# ADVANCED HIGHER

# 2010

[BLANK PAGE]

# X012/701

| NATIONAL QUALIFICATIONS 2010 | WEDNESDAY, 2 JUNE 9.00 AM – 11.30 AM | CHEMISTRY ADVANCED HIGHER |
|---|---|---|

Reference may be made to the Chemistry Higher and Advanced Higher Data Booklet.

**SECTION A – 40 marks**

Instructions for completion of **SECTION A** are given on page two.

For this section of the examination you must use an **HB pencil**.

**SECTION B – 60 marks**

All questions should be attempted.

**Answers must be written clearly and legibly in ink.**

## SECTION A

**Read carefully**

1 Check that the answer sheet provided is for **Chemistry Advanced Higher (Section A)**.

2 For this section of the examination you must use an **HB pencil** and, where necessary, an eraser.

3 Check that the answer sheet you have been given has **your name**, **date of birth**, **SCN** (Scottish Candidate Number) and **Centre Name** printed on it.

Do not change any of these details.

4 If any of this information is wrong, tell the Invigilator immediately.

5 If this information is correct, **print** your name and seat number in the boxes provided.

6 The answer to each question is **either** A, B, C or D. Decide what your answer is, then, using your pencil, put a horizontal line in the space provided (see sample question below).

7 There is **only one correct** answer to each question.

8 Any rough working should be done on the question paper or the rough working sheet, **not** on your answer sheet.

9 At the end of the examination, put the **answer sheet for Section A inside the front cover of your answer book**.

**Sample Question**

To show that the ink in a ball-pen consists of a mixture of dyes, the method of separation would be

A chromatography

B fractional distillation

C fractional crystallisation

D filtration.

The correct answer is **A**—chromatography. The answer **A** has been clearly marked in **pencil** with a horizontal line (see below).

**Changing an answer**

If you decide to change your answer, carefully erase your first answer and using your pencil, fill in the answer you want. The answer below has been changed to **D**.

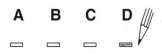

1. An atom of iron contains 26 electrons.

   Which of the following diagrams below correctly represents the distribution of electrons in the 3d and 4s orbitals in an atom of iron in its ground state?

   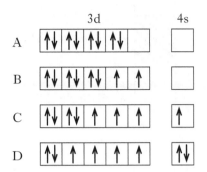

2. An atom has the electronic configuration

   $$1s^2 2s^2 2p^6 3s^2 3p^1$$

   What is the charge on the most probable ion formed by this element?

   A  +1

   B  +2

   C  +3

   D  +4

3. According to the aufbau principle, electrons fill orbitals in the order

   A  1s 2s 2p 3s 3p 4s 4p 3d

   B  1s 2s 2p 3s 3d 3p 4s 4p

   C  1s 2s 2p 3s 3p 3d 4s 4p

   D  1s 2s 2p 3s 3p 4s 3d 4p.

4. Which line in the graph represents the trend in successive ionisation energies of a Group 2 element?

   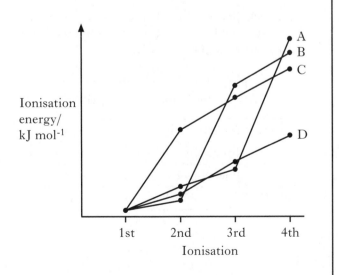

5. Which of the following statements about atomic emission spectroscopy is **incorrect**?

   A  Each element provides a characteristic spectrum.

   B  Visible light is used to promote electrons to higher energy levels.

   C  The lines arise from electron transitions between one energy level and another.

   D  The quantity of the element can be determined from the intensity of radiation transmitted.

6. Which of the following diagrams best represents the arrangement of atoms in the $IF_4^-$ ion?

   Note: a lone pair of electrons is represented by ••

   A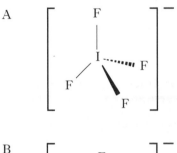

   B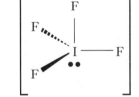

   C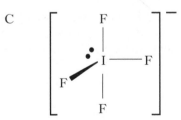

   D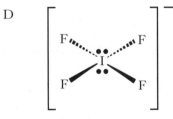

**[Turn over**

7. Which of the following solids is likely to have the same type of crystal lattice structure as caesium chloride?

   A   $Ba^{2+}O^{2-}$

   B   $Fe^{2+}O^{2-}$

   C   $Ag^{+}I^{-}$

   D   $Ni^{2+}O^{2-}$

8. An ionic hydride is added to water.

   Which line in the table correctly describes the gas produced and the type of solution formed?

   |   | Gas produced | Type of solution formed |
   |---|---|---|
   | A | hydrogen | acidic |
   | B | hydrogen | alkaline |
   | C | oxygen | acidic |
   | D | oxygen | alkaline |

9. An element forms an oxide which is a gas at room temperature.

   Which type of bonding is likely to be present in the element?

   A   Ionic

   B   Metallic

   C   Polar covalent

   D   Non-polar covalent

10. Which of the following oxides would produce the solution with the greatest conductivity when 0·1 mol is added to 250 cm$^3$ of water?

    A   $SO_2$

    B   $CO_2$

    C   $Na_2O$

    D   $Al_2O_3$

11. $2NH_3(g) \rightleftharpoons N_2(g) + 3H_2(g)$    $\Delta H^\circ = 92\,kJ\,mol^{-1}$

    The conditions favouring the decomposition of ammonia are

    A   low pressure and low temperature

    B   high pressure and low temperature

    C   low pressure and high temperature

    D   high pressure and high temperature.

12. An aqueous solution of iodine was shaken with cyclohexane until equilibrium was established.

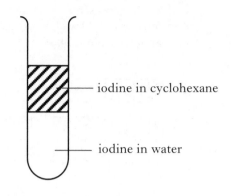

— iodine in cyclohexane

— iodine in water

Some solid iodine was added to the test tube and the contents shaken until equilibrium was re-established.

Which line in the table shows the effects caused by the addition of the solid iodine?

|   | Concentration of iodine in water | Concentration of iodine in cyclohexane | Partition coefficient |
|---|---|---|---|
| A | increases | increases | no change |
| B | increases | increases | increases |
| C | no change | increases | no change |
| D | increases | no change | increases |

13. An acid is a substance which

    A   donates a proton leaving a conjugate acid

    B   donates a proton leaving a conjugate base

    C   accepts a proton leaving a conjugate acid

    D   accepts a proton leaving a conjugate base.

14. The pH ranges over which some indicators change colour are shown below.

    Which line in the table shows the indicator most suitable for the titration of hydrochloric acid with ammonia solution?

    |   | Indicator | pH range |
    |---|---|---|
    | A | Methyl orange | 4·2 – 6·3 |
    | B | Bromothymol blue | 6·0 – 7·6 |
    | C | Phenol red | 6·8 – 8·4 |
    | D | Phenolphthalein | 8·3 – 10·0 |

**15.** The standard enthalpy of formation of magnesium bromide is the enthalpy change for the reaction

A  $Mg^{2+}(g) + 2Br^-(g) \rightarrow Mg^{2+}(Br^-)_2(s)$

B  $Mg^{2+}(g) + 2Br^-(g) \rightarrow Mg^{2+}(Br^-)_2(g)$

C  $Mg(s) + Br_2(g) \rightarrow Mg^{2+}(Br^-)_2(s)$

D  $Mg(s) + Br_2(\ell) \rightarrow Mg^{2+}(Br^-)_2(s)$.

**16.** The standard enthalpy of combustion of hydrogen is $-286$ kJ mol$^{-1}$.

The standard enthalpy of formation of water, in kJ mol$^{-1}$, is

A  $-286$

B  $-143$

C  $+143$

D  $+286$.

**17.** Which of the following enthalpy changes can be measured directly by experiment?

A  Bond enthalpy of C–H bond

B  Enthalpy of formation of ethane

C  Lattice enthalpy of magnesium oxide

D  Enthalpy of solution of potassium chloride

**18.** In which of the following does X represent the bond enthalpy for the O–H bond in water?

A  $H_2O(g) \rightarrow O(g) + H_2(g)$  $\Delta H = 2X$

B  $H_2O(g) \rightarrow O(g) + 2H(g)$  $\Delta H = 2X$

C  $H_2O(g) \rightarrow O(g) + H_2(g)$  $\Delta H = X$

D  $H_2O(g) \rightarrow O(g) + 2H(g)$  $\Delta H = X$

**19.** The enthalpy change for

$Li^+(g) + Br^-(g) \rightarrow Li^+(aq) + Br^-(aq)$

is

A  the enthalpy of formation of lithium bromide

B  the enthalpy of solution of lithium bromide

C  the sum of the hydration enthalpies of lithium and bromide ions

D  the sum of the first ionisation energy of lithium and the electron affinity of bromine.

Questions **20** and **21** refer to the Born-Haber cycle below.

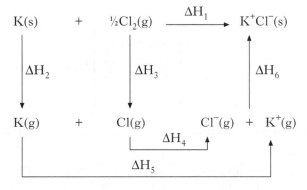

**20.** The enthalpy change which requires the input of most energy is

A  $\Delta H_2$

B  $\Delta H_3$

C  $\Delta H_4$

D  $\Delta H_5$.

**21.** The main enthalpy term which ensures that $\Delta H_1$ is exothermic is

A  $\Delta H_3$

B  $\Delta H_4$

C  $\Delta H_5$

D  $\Delta H_6$.

**22.** Which line in the table shows the correct signs of $\Delta G^\circ$ and $E^\circ$ for a feasible reaction occurring under standard conditions?

|   | $\Delta G^\circ$ | $E^\circ$ |
|---|---|---|
| A | + | + |
| B | + | − |
| C | − | + |
| D | − | − |

**23.** For the following cell

$Ni(s) \mid Ni^{2+}(aq) \parallel Cu^{2+}(aq) \mid Cu(s)$

the species being reduced is

A  $Ni(s)$

B  $Ni^{2+}(aq)$

C  $Cu^{2+}(aq)$

D  $Cu(s)$.

**24.**

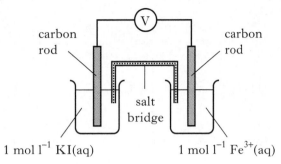

carbon rod    carbon rod

salt bridge

$1 \text{ mol l}^{-1} \text{ KI(aq)}$    $1 \text{ mol l}^{-1} \text{ Fe}^{3+}\text{(aq)}$

In the electrochemical cell shown above, operating under standard conditions, the emf produced would be

A   0·23 V

B   0·58 V

C   1·00 V

D   2·88 V.

**25.**   $C_3H_7Cl + C_2H_5O^- \rightarrow C_3H_7OC_2H_5 + Cl^-$

The above reaction is

A   an elimination reaction

B   a nucleophilic addition reaction

C   a nucleophilic substitution reaction

D   an electrophilic substitution reaction.

**26.**   Which of the following does **not** occur in the reaction between methane and chlorine?

A   A chain reaction

B   Homolytic fission

C   Free radical formation

D   An addition reaction

**27.**   Which of the following compounds is likely to be the most soluble in water?

A

$$H-\underset{\overset{|}{H}}{\overset{\overset{|}{H}}{C}}-\underset{\overset{|}{H}}{\overset{\overset{|}{H}}{C}}-\underset{\overset{\|}{O}}{C}-\underset{\overset{|}{H}}{\overset{\overset{|}{H}}{C}}-\underset{\overset{|}{H}}{\overset{\overset{|}{H}}{C}}-H$$

B

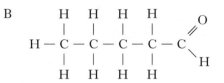

C

$$H-\underset{\overset{|}{H}}{\overset{\overset{|}{H}}{C}}-\underset{\overset{|}{H}}{\overset{\overset{|}{H}}{C}}-\underset{\overset{|}{H}}{\overset{\overset{|}{H}}{C}}-\underset{\overset{|}{H}}{\overset{\overset{|}{H}}{C}}-C\overset{\displaystyle O}{\underset{\displaystyle OH}{}}$$

D

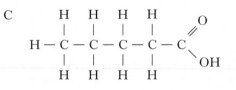

**28.**   The sideways overlap of two parallel atomic orbitals lying perpendicular to the axis of the bond is known as

A   hybridisation

B   a pi bond

C   a sigma bond

D   a double bond.

29. If the structure of 3-methylcyclobutene can be represented by

then the structure of 1-ethyl-3-methylcyclopentene will be represented by

A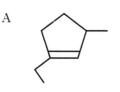

B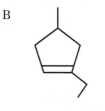

C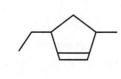

D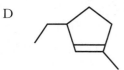

30. Caryophyllene ($C_{15}H_{24}$) is an unsaturated cyclic hydrocarbon.

Complete hydrogenation of caryophyllene gives a saturated hydrocarbon $C_{15}H_{28}$.

Which line in the table shows the correct numbers of double bonds and rings in caryophyllene?

| | Number of double bonds | Number of rings |
|---|---|---|
| A | 2 | 1 |
| B | 2 | 2 |
| C | 4 | 2 |
| D | 4 | 4 |

31. Which of the following is most reactive as a nucleophile?

A $Br_2$

B $CH_3I$

C $NH_4^+$

D $NH_3$

32. Hydrogen bonding occurs in

A $CH_3I$

B $CH_3OH$

C $CH_3OCH_3$

D $CH_3CH_2CHO$.

33. Cinnamaldehyde, which can be extracted from cinnamon, has the structure:

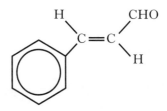

Cinnamaldehyde will **not** react with

A sodium metal

B bromine solution

C lithium aluminium hydride

D acidified potassium dichromate.

34. Which of the following compounds would be produced by passing ammonia gas into dilute ethanoic acid?

A $CH_3CONH_2$

B $CH_3COO^-NH_4^+$

C $NH_2CH_2COOH$

D $CH_3CH_2NH_3^+Cl^-$

[Turn over

**35.** Secondary amines react with carbonyl compounds to form unsaturated amines known as enamines as shown

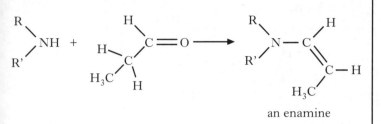

an enamine

Which carbonyl compound would react with $(CH_3)_2NH$ to form the enamine with the following structure?

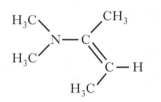

A  Propanal

B  Propanone

C  Butanal

D  Butanone

**36.**

⬡  $\xrightarrow{HNO_3/H_2SO_4}$  Product **X**

Which line in the table is correct for the reaction above?

|   | Type of reaction | Product **X** |
|---|---|---|
| A | electrophilic substitution | ⬡—NO₂ |
| B | electrophilic substitution | ⬡—SO₃H |
| C | nucleophilic substitution | ⬡—NO₂ |
| D | nucleophilic substitution | ⬡—SO₃H |

**37.** Which of the following statements about the benzene molecule is **not** true?

A  It is planar.

B  It has empirical formula CH.

C  It is readily attacked by bromine.

D  Its C—C bonds are equal in length.

**38.** Which of the following could **not** exist in isomeric forms?

A  $C_3H_6$

B  $C_3H_8$

C  $C_3H_7Br$

D  $C_2H_4Cl_2$

**39.** Which of the following causes the separation of the ions in a mass spectrometer?

A  A magnetic field

B  A vacuum pump

C  An ionisation chamber

D  Electron bombardment

**40.** Which of the following compounds is most likely to show an infra-red absorption at 2725 cm⁻¹?

A  $CH_3 - \overset{\overset{\displaystyle O}{\|}}{C} - CH_3$

B  $HOCH_2CH = CH_2$

C  $CH_3 - CH_2 - C\overset{\displaystyle O}{\underset{\displaystyle H}{\diagdown}}$

D  $CH_3 - O - CH = CH_2$

## SECTION B

**60 marks are available in this section of the paper.**

**All answers must be written clearly and legibly in ink.**

*Marks*

1. The first argon compound was prepared by shining light of wavelength 160 nm onto a mixture of argon and hydrogen fluoride at a temperature of 7·5 K. The hydrogen fluoride reacted with the argon to form HArF.

   (a) Calculate the energy, in kJ mol$^{-1}$, associated with light of wavelength 160 nm.      2

   (b) Supposing HArF is covalent,

      (i) predict the total number of electron pairs, bonding and non-bonding, which surround the Ar atom in the HArF molecule.      1

      (ii) what shape do the electron pairs around the Ar atom in an HArF molecule adopt?      1

      **(4)**

2. Complex ions **A** and **B** are isomeric and have the formula $[Cr(H_2O)_4Cl_2]^+$.

   (a) Calculate the oxidation number of chromium in the complex ion.      1

   (b) Name the complex ion.      1

   (c) The structural formula for complex ion **A** is

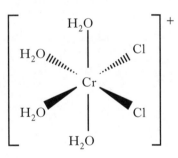

   Draw the structural formula for complex ion **B**.      1

      **(3)**

**[Turn over**

*Marks*

3.  The Thermit process can be used to extract iron from iron(III) oxide.

$$2Al(s) + Fe_2O_3(s) \rightarrow 2Fe(s) + Al_2O_3(s)$$

| Substance | Standard enthalpy of formation, $\Delta H°/kJ\ mol^{-1}$ | Standard entropy, $S°/J\ K^{-1}\ mol^{-1}$ |
|---|---|---|
| Al(s) | 0 | 28·0 |
| $Fe_2O_3(s)$ | -824 | 87·0 |
| Fe(s) | 0 | 27·0 |
| $Al_2O_3(s)$ | -1676 | 51·0 |

For the Thermit process, use the data in the table to calculate

(a)  the standard enthalpy change, $\Delta H°$                                                      1

(b)  the standard entropy change, $\Delta S°$                                                       1

(c)  the standard free energy change, $\Delta G°$.                                                   2

                                                                                                    (4)

*Marks*

4. In a PPA, a sample of steel was treated in a sequence of reactions to determine the manganese content.

$$Mn(s) \xrightarrow[\text{Step one}]{HNO_3} Mn^{2+}(aq) \xrightarrow[\text{Step two}]{KIO_4} MnO_4^-(aq)$$

The absorbance of a sample of the permanganate solution formed was analysed using a colorimeter fitted with a 520 nm filter. Optically matched cuvettes were used throughout.

(a)  (i)   Apart from the steel dissolving in the hot nitric acid, what would have been observed at step one?  1

 (ii)   In step two, $Mn^{2+}(aq)$ is converted into $MnO_4^-(aq)$. What is the role of the potassium periodate, $KIO_4$?  1

(iii)   Why was a 520 nm filter used?  1

(b)  A series of standard permanganate solutions were used to produce the calibration graph below.

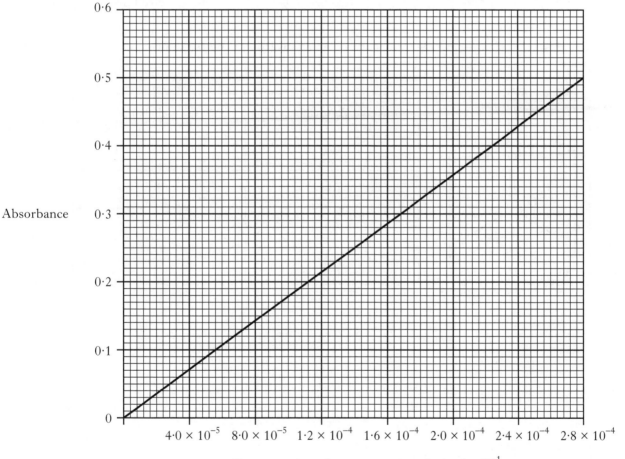

Concentration of permanganate solution/mol l$^{-1}$

The results of the experiment are shown below.

Mass of steel used                     =   0·19 g

Absorbance of permanganate solution    =   0·25

Total volume of permanganate solution  =   100 cm$^3$

Use the graph and the results to calculate the percentage, by mass, of manganese in the sample of steel.  3

**(6)**

*Marks*

5. The nitrate ion has three equivalent resonance structures. One of these structures is shown below.

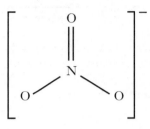

(a) Draw a similar diagram to show one of the other two resonance structures.          1

(b) The formal charge on an atom in a resonance structure can be found using the expression

$$\text{Formal charge} = \left(\begin{array}{c}\text{Group in} \\ \text{Periodic Table}\end{array}\right) - \left(\begin{array}{c}\text{Number of lone pair} \\ \text{electrons}\end{array}\right) - \tfrac{1}{2}\left(\begin{array}{c}\text{Number of bonding} \\ \text{electrons}\end{array}\right).$$

Use this expression to find the formal charge on atoms (b), (c) and (d) shown in the table below.

| Resonance structure | Atom | Formal charge |
|---|---|---|
| | (a) | +1 |
| | (b) | ? |
| | (c) | ? |
| | (d) | ? |

2

(3)

*Marks*

6. The formula of potassium hydrogen oxalate can be written as $K_xH_y(C_2O_4)_z$.

   In an experiment to determine the values of **x**, **y** and **z**, 4·49 g of this compound was dissolved in water and the solution made up to one litre.

   (a) 20·0 cm³ of the solution was pipetted into a conical flask and then titrated with 0·0200 mol l⁻¹ acidified potassium permanganate at 60 °C. The average titre volume was 16·5 cm³.

   The equation for the reaction taking place in the conical flask is

   $$5C_2O_4^{2-} + 16H^+ + 2MnO_4^- \rightarrow 2Mn^{2+} + 10CO_2 + 8H_2O$$

   (i) What colour change would indicate the end point of the titration?    1

   (ii) From the titration result, calculate the number of moles of oxalate ions, $C_2O_4^{2-}$, in 20·0 cm³ of the solution.    1

   (iii) Calculate the mass of oxalate ions in one litre of the solution.    1

   (iv) Using another analytical procedure, 4·49 g of potassium hydrogen oxalate was found to contain 0·060 g of hydrogen.

   Use this information with the answer to (a)(iii) to calculate the mass of potassium in this sample.    1

   (b) Calculate the values of **x**, **y** and **z**.    2

   (6)

7. The expression for the equilibrium constant of an esterification reaction is

   $$K = \frac{[CH_3COOCH_2CH_3]\,[H_2O]}{[CH_3COOH]\,[CH_3CH_2OH]}$$

   (a) Write the chemical equation for this esterification reaction.    1

   (b) In an experiment to determine the value of the equilibrium constant, 0·70 moles of ethanoic acid and 0·68 moles of ethanol were mixed in a conical flask. The flask was stoppered to prevent the contents escaping and then placed in a water bath at 50 °C.

   At equilibrium the mixture contained 0·24 moles of ethanoic acid.

   (i) Why is it important to prevent the contents of the flask escaping?    1

   (ii) Calculate K at 50 °C.    3

   (5)

8. Nicotinic acid is used in the treatment of high cholesterol levels. A structural formula for nicotinic acid is

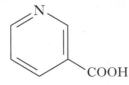

   (a) Write an equation to show the dissociation of nicotinic acid in water.    1

   (b) The $K_a$ value of nicotinic acid is $1·4 \times 10^{-5}$.

   Calculate the concentration of a nicotinic acid solution which has a pH of 3·77.    3

   (4)

*Marks*

9. The rate equation for the reaction between nitrogen dioxide and fluorine is

$$Rate = k[NO_2][F_2]$$

A proposed reaction mechanism is

Step one    $NO_2 + F_2 \rightarrow NO_2F + F$

Step two    $NO_2 + F \rightarrow NO_2F.$

(a) Which step in the proposed reaction mechanism would be **faster**?     1

(b) Write a balanced equation for the overall reaction.     1

(c) What is the overall order of the reaction?     1

(d)

| Experiment | $[NO_2]$/mol l$^{-1}$ | $[F_2]$/mol l$^{-1}$ | Initial rate/ mol l$^{-1}$ s$^{-1}$ |
|---|---|---|---|
| 1 | 0·001 | 0·003 | $1\cdot2 \times 10^{-4}$ |
| 2 | 0·006 | 0·001 | $2\cdot4 \times 10^{-4}$ |
| 3 | 0·002 | 0·004 | $3\cdot2 \times 10^{-4}$ |

Use the data in the table to calculate a value for the rate constant, k, including the appropriate units.     2

**(5)**

10. Alkenes can be prepared from alcohols.

In a PPA, 22·56 g of cyclohexanol was dehydrated using an excess of concentrated phosphoric acid. The reaction mixture was then distilled. The crude cyclohexene was added to a separating funnel containing a solution which was used to wash the cyclohexene and improve the separation of the aqueous and organic layers. The organic layer was separated and treated with anhydrous calcium chloride before it was distilled to yield 6·52 g of pure cyclohexene.

(a) Why was concentrated phosphoric acid used as the dehydrating agent rather than concentrated sulphuric acid?     1

(b) Name the solution that the crude cyclohexene was added to in the separating funnel.     1

(c) What was the function of the anhydrous calcium chloride?     1

(d) The relative formula masses of cyclohexanol and cyclohexene are 100 and 82 respectively.

Calculate the percentage yield of cyclohexene.     2

**(5)**

*Marks*

11.  Consider the following reaction scheme.

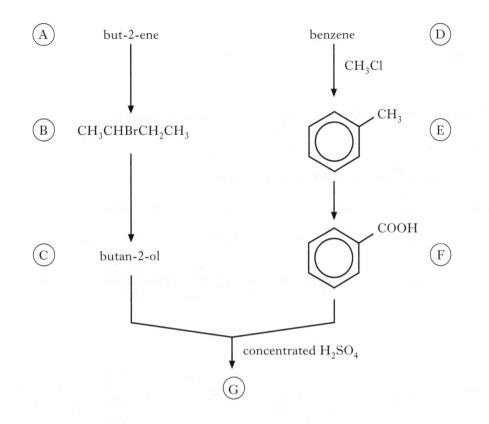

(a)  Explain why but-2-ene exhibits geometric isomerism yet its structural isomer but-1-ene does not.  **1**

(b)  But-2-ene undergoes electrophilic addition to form Ⓑ.

Draw a structure for the carbocation intermediate formed in this electrophilic addition reaction.  **1**

(c)  Name a reagent used to convert Ⓑ to Ⓒ.  **1**

(d)  Name a catalyst required in converting Ⓓ to Ⓔ.  **1**

(e)  Draw a structural formula for ester Ⓖ.  **1**

**(5)**

**[Turn over**

*Marks*

**12.** Consider the following reaction sequence.

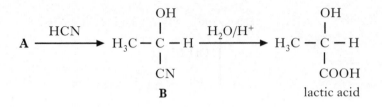

(*a*) Name compound **A**.  1

(*b*) To which class of organic compounds does compound **B** belong?  1

(*c*) Name the type of reaction taking place in converting compound **B** into lactic acid.  1

(*d*) Lactic acid in the form of lactate ions is dehydrogenated in the liver by the enzyme, lactate dehydrogenase.

The diagram shows how one of the optical isomers of the lactate ion binds to an active site of lactate dehydrogenase.

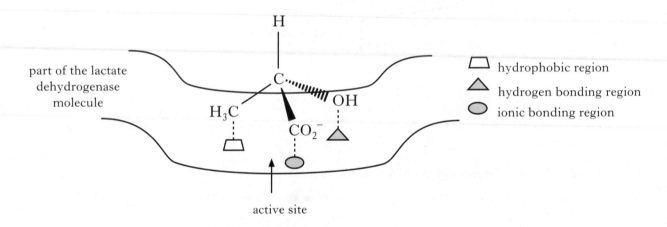

(i) Which type of intermolecular force is involved when the methyl group of the lactate ion binds to the hydrophobic region of the active site?  1

(ii) Draw a structure for the other optical isomer of the lactate ion.  1

(iii) Explain why this other optical isomer of the lactate ion cannot bind as efficiently to the active site of lactate dehydrogenase.  1

**(6)**

*Marks*

**13.** Compound **A** has molecular formula $C_4H_{10}O$.

(a) To which two classes of organic compounds could **A** belong? 2

(b) Compound **A** reacts with acidified potassium dichromate solution to form **B** which has molecular formula $C_4H_8O$.

The proton nmr spectrum of **B** shows three peaks. Analysis of this spectrum produces the following data.

| Peak | Chemical shift/ppm | Relative area under peak |
|------|--------------------|--------------------------|
| 1 | 0·95 | 3 |
| 2 | 2·05 | 3 |
| 3 | 2·35 | 2 |

Considering all the evidence above:

(i) draw a structural formula for **B**; 1

(ii) name **A**. 1

(4)

[*END OF QUESTION PAPER*]

[BLANK PAGE]

[BLANK PAGE]

# X012/701

| NATIONAL QUALIFICATIONS 2011 | THURSDAY, 26 MAY 9.00 AM – 11.30 AM | CHEMISTRY ADVANCED HIGHER |
|---|---|---|

Reference may be made to the Chemistry Higher and Advanced Higher Data Booklet .

**SECTION A – 40 marks**

Instructions for completion of **SECTION A** are given on page two.

For this section of the examination you must use an **HB pencil**.

**SECTION B – 60 marks**

All questions should be attempted.

**Answers must be written clearly and legibly in ink.**

## SECTION A

### Read carefully

1  Check that the answer sheet provided is for **Chemistry Advanced Higher (Section A)**.

2  For this section of the examination you must use an **HB pencil** and, where necessary, an eraser.

3  Check that the answer sheet you have been given has **your name**, **date of birth**, **SCN** (Scottish Candidate Number) and **Centre Name** printed on it.

Do not change any of these details.

4  If any of this information is wrong, tell the Invigilator immediately.

5  If this information is correct, **print** your name and seat number in the boxes provided.

6  The answer to each question is **either** A, B, C or D.  Decide what your answer is, then, using your pencil, put a horizontal line in the space provided (see sample question below).

7  There is **only one correct** answer to each question.

8  Any rough working should be done on the question paper or the rough working sheet, **not** on your answer sheet.

9  At the end of the exam, put the **answer sheet for Section A inside the front cover of your answer book**.

### Sample Question

To show that the ink in a ball-pen consists of a mixture of dyes, the method of separation would be

A  chromatography

B  fractional distillation

C  fractional crystallisation

D  filtration.

The correct answer is **A**—chromatography.  The answer **A** has been clearly marked in **pencil** with a horizontal line (see below).

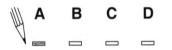

### Changing an answer

If you decide to change your answer, carefully erase your first answer and using your pencil, fill in the answer you want.  The answer below has been changed to **D**.

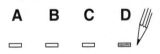

1. Which of the following lines on the graph represents the trend in successive ionisation energies of a Group 3 element?

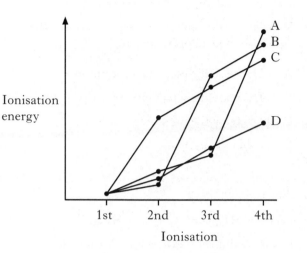

2. In colorimetry, as the concentration of a coloured solution decreases

A   the absorbance increases

B   the absorbance decreases

C   the radiation wavelength increases

D   the radiation wavelength decreases.

3. Which of the following molecules has the greatest number of non-bonding electron pairs (lone pairs)?

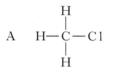

A

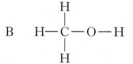

B

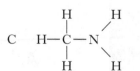

C

D   H—C＝O
        |
        H

4. What is the change in the three-dimensional arrangement of the bonds round the P atom in the following reaction?

$$PF_5 \rightarrow PF_3 + F_2$$

A   Tetrahedral to pyramidal

B   Octahedral to trigonal planar

C   Trigonal bipyramidal to pyramidal

D   Trigonal bipyramidal to trigonal planar

5. The ratio of the ionic radii in sodium chloride is approximately 1:2, whereas in caesium chloride it is approximately 1:1. A compound XY contains $X^+$ ions with a radius of 133 pm and $Y^-$ ions with a radius of 220 pm.

In a crystal of XY, how many $Y^-$ ions surround each $X^+$ ion as its nearest neighbour?

A   1

B   2

C   6

D   8

6. An example of a p-type semiconductor is silicon doped with

A   carbon

B   arsenic

C   aluminium

D   phosphorus.

7. Which of the following solid oxides would **not** lower the pH when added to sodium hydroxide solution?

A   $Li_2O$

B   $SiO_2$

C   $P_4O_{10}$

D   $Al_2O_3$

**[Turn over**

8. Which of the following is least likely to produce fumes of hydrogen chloride when added to water?

A $PCl_5$

B $SiCl_4$

C $AlCl_3$

D $MgCl_2$

9. A white solid gives an orange-yellow flame colour. When added to water, hydrogen gas is released and an alkaline solution is formed.

The solid could be

A sodium oxide

B calcium oxide

C sodium hydride

D calcium hydride.

10. Which of the following ions is **least** likely to be coloured?

A $Ti(H_2O)_6^{3+}$

B $Cr(NH_3)_6^{3+}$

C $Ni(H_2O)_6^{2+}$

D $Zn(NH_3)_4^{2+}$

11. What volume of $0 \cdot 25 \, mol \, l^{-1}$ calcium nitrate is required to make, by dilution with water, $500 \, cm^3$ of a solution with a **nitrate** ion concentration of $0 \cdot 1 \, mol \, l^{-1}$?

A $50 \, cm^3$

B $100 \, cm^3$

C $200 \, cm^3$

D $400 \, cm^3$

12. Hydrogen for use in ammonia production is produced by the endothermic reaction:

$$CH_4(g) + H_2O(g) \rightleftharpoons CO(g) + 3H_2(g)$$

Which of the following will increase the equilibrium yield of hydrogen?

A Decrease the methane concentration

B Decrease the temperature

C Decrease the pressure

D Add a catalyst

13. The reaction

$$CO(g) + 3H_2(g) \rightleftharpoons CH_4(g) + H_2O(g)$$

has an equilibrium constant of $3 \cdot 9$ at $950 \, °C$.

The equilibrium concentrations of $CO(g)$, $H_2(g)$ and $H_2O(g)$ are given in the table.

| Substance | Equilibrium concentration/$mol \, l^{-1}$ |
|---|---|
| $CO(g)$ | $0 \cdot 500$ |
| $H_2(g)$ | $0 \cdot 100$ |
| $H_2O(g)$ | $0 \cdot 040$ |

What is the equilibrium concentration of $CH_4(g)$, in $mol \, l^{-1}$, at $950 \, °C$?

A $0 \cdot 049$

B $0 \cdot 200$

C $4 \cdot 90$

D $20 \cdot 0$

14.

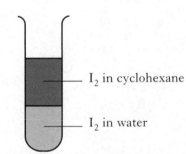

I$_2$ in cyclohexane

I$_2$ in water

The partition coefficient for the above system can be altered by

A adding more iodine

B adding more cyclohexane

C changing the temperature

D shaking the mixture thoroughly.

**15.** Gas liquid chromatography could be used to separate a mixture of hydrocarbons. The mixture is passed through a column packed with silica particles coated in a non-polar liquid. Helium can be used to carry the mixture through the column.

Which line in the table identifies correctly the stationary and mobile phases in this chromatographic separation?

|   | Stationary phase | Mobile phase |
|---|---|---|
| A | silica | helium |
| B | silica | non-polar liquid |
| C | non-polar liquid | helium |
| D | non-polar liquid | hydrocarbon mixture |

**16.** Under certain conditions liquid ammonia ionises as shown:

$$2NH_3 \rightleftharpoons NH_4^+ + NH_2^-$$

Which line in the table shows the correct conjugate acid and conjugate base for this ionisation?

|   | Conjugate acid | Conjugate base |
|---|---|---|
| A | $NH_3$ | $NH_4^+$ |
| B | $NH_4^+$ | $NH_3$ |
| C | $NH_2^-$ | $NH_4^+$ |
| D | $NH_4^+$ | $NH_2^-$ |

**17.** The activation energies for the reactions

(1)   $H_2(g) + I_2(g) \rightarrow 2HI(g)$

(2)   $2HI(g) \rightarrow H_2(g) + I_2(g)$

are 165 kJ and 179 kJ respectively. The enthalpy change for reaction (2) is

A    −14 kJ

B    +14 kJ

C    −344 kJ

D    +344 kJ.

**18.** The standard enthalpy of formation of strontium chloride is the enthalpy change for which of the following reactions?

A    $Sr(g) + Cl_2(g) \rightarrow SrCl_2(s)$

B    $Sr(s) + Cl_2(g) \rightarrow SrCl_2(s)$

C    $Sr^{2+}(g) + 2Cl^-(g) \rightarrow SrCl_2(s)$

D    $Sr^{2+}(aq) + 2Cl^-(aq) \rightarrow SrCl_2(s)$

**19.** Consider the following thermochemical cycle which is not drawn to scale.

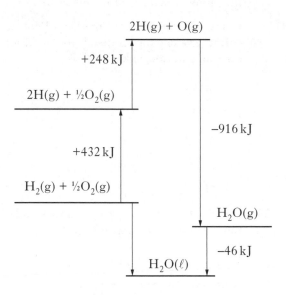

The enthalpy change for the reaction

$2H_2(g) + O_2(g) \rightarrow 2H_2O(\ell)$ is

A    −564 kJ

B    −282 kJ

C    +564 kJ

D    +1642 kJ.

20. In the presence of bright light, hydrogen and chlorine react explosively. One step in the reaction is shown below.

$H_2(g) + Cl(g) \rightarrow HCl(g) + H(g)$

The enthalpy change for this step can be represented as the bond enthalpy of

A    $(H—H) + (Cl—Cl)$

B    $(H—H) - (Cl—Cl)$

C    $(H—H) + (H—Cl)$

D    $(H—H) - (H—Cl)$.

21. The standard enthalpy of atomisation of bromine is the enthalpy change for the reaction

A    $\frac{1}{2}Br_2(s) \rightarrow Br(g)$

B    $\frac{1}{2}Br_2(\ell) \rightarrow Br(g)$

C    $\frac{1}{2}Br_2(g) \rightarrow Br(g)$

D    $Br_2(g) \rightarrow 2Br(g)$.

22. The enthalpy of solution of a compound can be calculated from its lattice enthalpy and the hydration enthalpies of its ions.

Using information from the Data Booklet, the correct value for enthalpy of solution of calcium chloride, in $kJ\ mol^{-1}$, is

A    $-155$

B    $+155$

C    $-209$

D    $+209$.

23. Which of the following reactions would show the greatest decrease in entropy?

A    $H_2(g) + F_2(g) \rightarrow 2HF(g)$

B    $KNO_3(s) \rightarrow KNO_2(s) + \frac{1}{2}O_2(g)$

C    $CO_3^{2-}(aq) + 2H^+(aq) \rightarrow H_2O(\ell) + CO_2(g)$

D    $CO_3^{2-}(aq) + CO_2(g) + H_2O(\ell) \rightarrow 2HCO_3^-(aq)$

24. Which of the following alcohols would have the greatest entropy at $90\,^\circ C$?

A    Propan-1-ol

B    Propan-2-ol

C    Butan-1-ol

D    Butan-2-ol

25. Which of the following redox equations represents a reaction which is not feasible under standard conditions?

A    $F_2(g) + 2Cl^-(aq) \rightarrow 2F^-(aq) + Cl_2(g)$

B    $Cl_2(g) + 2Br^-(aq) \rightarrow 2Cl^-(aq) + Br_2(\ell)$

C    $F_2(g) + 2Br^-(aq) \rightarrow 2F^-(aq) + Br_2(\ell)$

D    $I_2(s) + 2Br^-(aq) \rightarrow 2I^-(aq) + Br_2(\ell)$

26. Propene can be produced by heating 1-bromopropane with ethanolic potassium hydroxide.

This reaction is an example of

A    reduction

B    hydrolysis

C    elimination

D    condensation.

**27.** The structures of three alcohols, **P**, **Q**, and **R** are shown.

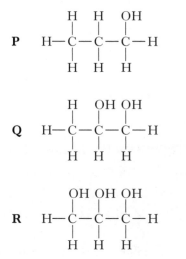

Which line in the table describes correctly the trends in boiling points and viscosities on moving from **P** to **Q** to **R**?

|   | Boiling point | Viscosity |
|---|---|---|
| A | increases | increases |
| B | increases | decreases |
| C | decreases | increases |
| D | decreases | decreases |

**28.** Which of the following best describes the bonding in ethane?

A    $sp^2$ hybridisation of the carbon atoms giving sigma bonds only

B    $sp^2$ hybridisation of the carbon atoms giving sigma and pi bonds

C    $sp^3$ hybridisation of the carbon atoms giving sigma bonds only

D    $sp^3$ hybridisation of the carbon atoms giving sigma and pi bonds

**29.** Part of a possible chain reaction mechanism for chlorine reacting with methane is:

$Cl_2 \rightarrow 2Cl\bullet$

$Cl\bullet + CH_4 \rightarrow HCl + CH_3\bullet$

$CH_3\bullet + Cl_2 \rightarrow CH_3Cl + Cl\bullet$

Which of the following will **not** be a termination step in this reaction?

A    $H\bullet + Cl\bullet \rightarrow HCl$

B    $Cl\bullet + Cl\bullet \rightarrow Cl_2$

C    $CH_3\bullet + CH_3\bullet \rightarrow C_2H_6$

D    $CH_3\bullet + Cl\bullet \rightarrow CH_3Cl$

**30.** Pyridine, $C_5H_5N$, has the following structure:

Which line in the table shows the correct numbers of $\sigma$ and $\pi$ bonds in a molecule of pyridine?

|   | Number of $\sigma$ bonds | Number of $\pi$ bonds |
|---|---|---|
| A | 3 | 11 |
| B | 6 | 3 |
| C | 11 | 3 |
| D | 12 | 3 |

**31.** The major product in the reaction of HCl with 2-methylpent-2-ene,

$$\underset{}{H_3C - \overset{\overset{\displaystyle CH_3}{|}}{C} = CH - CH_2 - CH_3} \text{ is}$$

A    2-chloro-2-methylpentane

B    3-chloro-2-methylpentane

C    2,3-dichloro-2-methylpentane

D    4-chloro-4-methylpentane.

**[Turn over**

32. A compound, **X**, reacts with the product of its own oxidation to form an ester.

    **X** could be

    A    propanal

    B    propan-1-ol

    C    propan-2-ol

    D    propanoic acid.

33. Which of the following amines does **not** have hydrogen bonds between its molecules in the liquid state?

    A    $CH_3CH_2CH_2CH_2NH_2$

    B    $CH_3CH_2NHCH_2CH_3$

    C    $(CH_3)_2CHCH_2NH_2$

    D    $(CH_3)_2NCH_2CH_3$

34. 1 mole of which of the following compounds would react with the largest volume of $1\,mol\,l^{-1}$ hydrochloric acid?

    A    $CH_3NHCH_3$

    B    $H_2NCH_2NH_2$

    C    $CH_2OHCHOHCH_2OH$

    D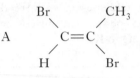

35. The conversion of benzene to monochlorobenzene using $Cl_2/FeCl_3$ involves

    A    nucleophilic addition

    B    nucleophilic substitution

    C    electrophilic addition

    D    electrophilic substitution.

36.

    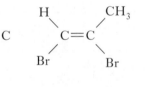

    Which species initially attacks the benzene molecule in the above reaction?

    A    $NO_3^-$

    B    $NO_2^+$

    C    $HSO_4^-$

    D    $NO_2$

37. Which of the following is the geometric isomer of *trans*-1,2-dibromopropene?

    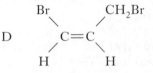

38. The mass spectrum of an organic compound, empirical formula $C_2H_4O$, shows a peak for the parent ion at mass/charge ratio of 88.

    The organic compound could **not** be

    A    ethanal

    B    butanoic acid

    C    ethyl ethanoate

    D    methyl propanoate.

**39.** From which region of the electromagnetic spectrum is energy absorbed in the production of proton nmr spectra?

A    X-rays

B    Visible

C    Infra-red

D    Radio waves

**40.** A compound, which has molecular formula $C_4H_8O$, has only 2 peaks in its low resolution proton nmr spectrum.

A possible structural formula for this compound is

A    $CH_3CH_2CCH_3$
         ‖
         O

B    $CH_3CH_2CH_2CHO$

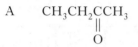

C

$CH_3$
|
$CH_3 — C — CH_3$
|
OH

D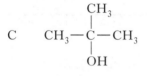

$H_2C — CH_2$
   ╱          ╲
$H_2C$      $CH_2$
    ╲    O   ╱

*[END OF SECTION A]*

**Candidates are reminded that the answer sheet for Section A MUST be placed INSIDE the front cover of your answer book.**

**[Turn over for SECTION B on *Page ten***

**SECTION B**

*Marks*

**60 marks are available in this section of the paper.**

**All answers must be written clearly and legibly in ink.**

1.  The compound, $Sn_2Ba_2(Sr_{0.5}Y_{0.5})Cu_3O_8$, has zero electrical resistance at 85 K.

    (*a*)  What name is given to this phenomenon?                                                         1

    (*b*)  Which liquid coolant can be used economically and safely at this temperature?                  1

                                                                                                         (2)

2.  When hydrogen is subjected to a high voltage in a gas discharge tube and the emitted light is passed
    through a prism the atomic emission spectrum produced is as shown below.

    hydrogen emission spectrum (visible region)

    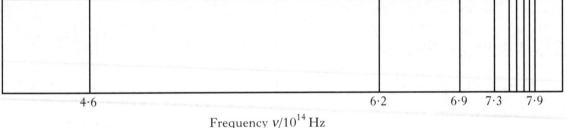

    Frequency $v/10^{14}$ Hz

    (*a*)  Which line in the spectrum is red?                                                             1

    (*b*)  The ionisation energy of hydrogen has a value of 1311 kJ mol$^{-1}$.

        (i)  Write the equation for the ionisation energy of hydrogen.                                    1

        (ii)  Calculate the wavelength of the light corresponding to this ionisation energy.              3

                                                                                                         (5)

*Marks*

3.  When a mixture of nitrogen monoxide and nitrogen dioxide is cooled to $-20\,^{\circ}C$ they react to form the clear blue liquid, dinitrogen trioxide.

$$NO + NO_2 \rightarrow N_2O_3$$

(a)  The oxidation state of nitrogen is **different** in each of these three compounds.

Calculate the oxidation states of the nitrogen in NO and $NO_2$ respectively.          1

(b)  Dinitrogen trioxide neutralises aqueous sodium hydroxide forming sodium nitrite and water.

The nitrite ion, $NO_2^-$, can be represented by two resonance structures.

One of these is

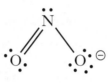

Draw the other resonance structure.          1

(c)  In aqueous solution the nitrite ion can be oxidised to the nitrate ion.

Write the ion-electron equation for this oxidation.          1

          (3)

**[Turn over**

*Marks*

**4.** Iron and manganese are transition metals which have many uses in industry.

The electronic configuration for iron, in its ground state, is

$$1s^2 2s^2 2p^6 3s^2 3p^6 3d^6 4s^2$$

(a) In terms of s, p and d orbitals write down the electronic configurations of

    (i) $Fe^{3+}$

    (ii) $Mn^{3+}$

in their ground states.　　2

    (iii) Explain why the $Fe^{3+}$ ion is more stable than the $Mn^{3+}$ ion.　　1

(b) The transition metal titanium is the seventh most abundant element in the Earth's crust.

Two of the reactions involved in the conversion of the ore ilmenite, $FeTiO_3$, into metallic titanium are shown below.

**Step 1**—Ilmenite is reacted with concentrated sulphuric acid.

$$FeTiO_3(s) + 3H_2SO_4(\ell) \rightarrow FeSO_4(aq) + Ti(SO_4)_2(aq) + 3H_2O(\ell)$$

**Step 2**—After separation the titanium sulphate is reacted with sodium hydroxide.

$$Ti(SO_4)_2(aq) + 4NaOH(aq) \rightarrow TiO_2(s) + 2H_2O(\ell) + 2Na_2SO_4(aq)$$

How many kilograms of titanium oxide can theoretically be produced from 3·25 kg of ilmenite?　　2

(c) Transition metals can form a wide variety of complexes. One such complex is ammonium tetrachlorocuprate(II).

Write the formula for this complex.　　1

　　(6)

*Marks*

5.   The PPA "Complexometric Determination of Nickel using EDTA" has two main stages.

   **Stage 1**    Preparation of nickel(II) sulphate solution.

   **Stage 2**    Titration of the nickel(II) sulphate solution with EDTA.

   The instructions for **Stage 1** are shown below.

   1.   Accurately weigh out approximately 2·6 g of hydrated nickel(II) sulphate, $NiSO_4.6H_2O$.

   2.   Transfer the hydrated nickel salt to a $100\,cm^3$ beaker, add $25\,cm^3$ of deionised water and stir to dissolve the solid.

   3.   Transfer the solution to a $100\,cm^3$ standard flask.

   4.

   5.

   6.   Stopper the flask and invert it several times to ensure the contents are thoroughly mixed.

   (*a*)   Complete the instructions for steps 4 and 5.          1

   (*b*)   In **Stage 2**, $25·0\,cm^3$ of the nickel(II) sulphate solution were titrated against $0·110\ mol\,l^{-1}$ EDTA solution.

   The results of the titrations are shown below

   |  | Rough titre | 1st titre | 2nd titre |
   |---|---|---|---|
   | Initial burette reading/$cm^3$ | 2·00 | 25·90 | 10·00 |
   | Final burette reading/$cm^3$ | 25·90 | 49·40 | 33·60 |
   | Volume of EDTA added/$cm^3$ | 23·90 | 23·50 | 23·60 |

   The equation for the reaction is represented by

   $$Ni^{2+}(aq) + [EDTA]^{4-}(aq) \rightarrow Ni[EDTA]^{2-}(aq)$$

   (i)    Name the indicator used to detect the end-point of the titration in this PPA.          1

   (ii)   EDTA acts as a hexadentate ligand. What shape is the complex ion $Ni[EDTA]^{2-}$?          1

   (iii)  The accurate mass of the nickel(II) sulphate used was 2·656 g.

   Calculate the percentage by mass of nickel present in the hydrated salt from these experimental results.          3

   **(6)**

**[Turn over**

*Marks*

6. The standard free energy change for a chemical reaction is given by the expression

$$\Delta G^\circ = \Delta H^\circ - T\Delta S^\circ$$

The expression can be rearranged to give

$$\Delta G^\circ = -\Delta S^\circ T + \Delta H^\circ$$

Plotting values of $\Delta G^\circ$ against T will therefore produce a straight line with gradient equal to $-\Delta S^\circ$.

The graph shows how $\Delta G^\circ$ varies with temperature for a particular chemical reaction.

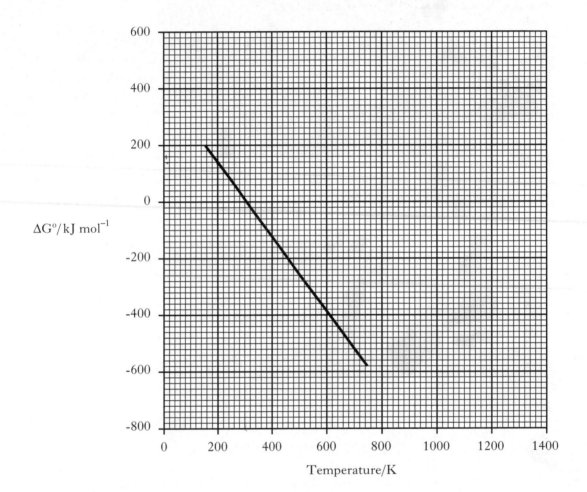

Use the graph to

(a) deduce the temperature at which the reaction just becomes feasible under standard conditions    **1**

(b) estimate the value of $\Delta H^\circ$, in kJ mol$^{-1}$, for the reaction    **1**

(c) calculate the value of $\Delta S^\circ$, in J K$^{-1}$ mol$^{-1}$.    **2**

**(4)**

*Marks*

7.  Consider the three reactions and their rate equations

    **Reaction** (1)    $2N_2O_5 \rightarrow 4NO_2 + O_2$    Rate = $k[N_2O_5]$

    **Reaction** (2)    $2NO + Cl_2 \rightarrow 2NOCl$    Rate = $k[NO]^2[Cl_2]$

    **Reaction** (3)    $2NH_3 \rightarrow N_2 + 3H_2$    Rate = $k[NH_3]^0$

    (a)  What is the overall order of Reaction (2)?    **1**

    (b)  The graph below was plotted using experimental results from one of the reactions.

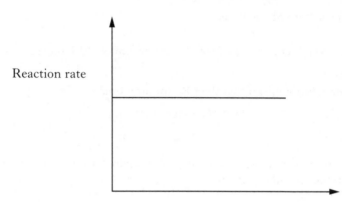

    Explain which of the reactions would give this graph.    **1**

    (c)  For Reaction (2), when the concentrations of NO and $Cl_2$ are both $0.250$ $mol\,l^{-1}$, the initial reaction rate is $1.43 \times 10^{-6}$ $mol\,l^{-1}\,s^{-1}$.

    Use this information to calculate the rate constant, k, including the appropriate units.    **2**

    **(4)**

8.  The reaction between hydrogen peroxide and potassium bromide is used to generate bromine to disinfect water supplies.

    The ion-electron equations involved in this reaction are

    $$Br_2(\ell) + 2e^- \rightarrow 2Br^-(aq) \qquad\qquad E^\circ = 1.07\,V$$

    $$H_2O_2(aq) + 2H^+(aq) + 2e^- \rightarrow 2H_2O(\ell) \qquad\qquad E^\circ = 1.77\,V$$

    (a)  Write the redox equation for the reaction.    **1**

    (b)  Calculate the standard free energy change, in $kJ\,mol^{-1}$, for this reaction.    **3**

    **(4)**

**[Turn over**

*Marks*

9. Buffer solutions are important in human biochemistry.

   (a) What is meant by a "buffer solution"?                                     1

   (b) Suggest the name of a salt which could be mixed with propanoic acid to prepare an acidic buffer solution.                                              1

   (c) The pH of an alkaline buffer solution can be found using the formula

$$pH = pK_w - pK_b + \log \frac{[base]}{[salt]}$$

   where   $K_w$ is the ionic product of water

   and     $K_b$ is the dissociation constant of the base.

   $1.05$ g of ammonium nitrate, $NH_4NO_3$, is dissolved in $100$ cm$^3$ of a $0.15$ mol l$^{-1}$ ammonia solution at $25\,°C$.

   Calculate the pH of this buffer solution given that the $pK_b$ for ammonia is $4.76$.      3

                                                                                  **(5)**

10. Chemists are developing compounds which block the ability of certain bacteria to bind to the surface of cells. This will help stop the spread of infection.

   (a) What name is given to the structural fragment of this type of compound which binds to a receptor?                                                        1

   (b) The diagram shows the structure of four of these compounds.

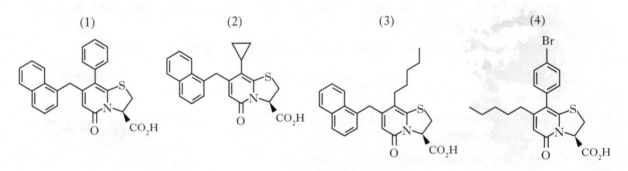

   Draw the structural fragment which is common to these compounds which allows them to bind to the relevant receptor.                                          1

                                                                                  **(2)**

*Marks*

11.  Meldrum's acid is a chemical named after the Scotsman, Andrew N. Meldrum who was the first to produce it.

Microanalysis showed that Meldrum's acid has a composition, by mass, of 50% C, 5·6% H, 44·4% O.

(a)  Use the percentage composition to calculate the empirical formula of Meldrum's acid.

**(Working must be shown)**                                                      1

(b)  Meldrum initially thought the structure was

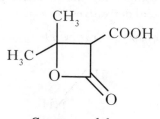

**Compound A**

The structure was later shown to be the isomer of **A** shown below.

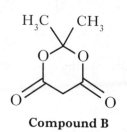

**Compound B**

(i)  What is the molecular formula of **A** and **B**?                            1

(ii)  The infra-red spectrum of isomer **A** would show a strong absorbance not shown by isomer **B**.

Identify the wave number range, in cm$^{-1}$, where this absorbance occurs.       1

(3)

**[Turn over**

*Marks*

12. Cinnamaldehyde is an aromatic compound found in cinnamon. It can also be prepared by the reaction of benzaldehyde and ethanal.

$$C_6H_5CHO + CH_3CHO \rightarrow C_6H_5CHCHCHO$$

(a) What type of reaction is this?      1

(b) Draw a full structural formula for cinnamaldehyde.      1

(c) All three of the carbonyl compounds shown above react with 2,4-dinitrophenylhydrazine, (Brady's reagent), forming solid derivatives.

The structure of 2,4-dinitrophenylhydrazine is

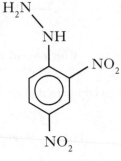

(i) Draw a structural formula of the compound formed when ethanal reacts with 2,4-dinitrophenylhydrazine.      1

(ii) The compound formed is impure.

How would this compound be purified?      1

(iii) How would the purified compound be used to show that the original carbonyl compound was ethanal?      1

(iv) 2,4-Dinitrophenylhydrazone derivatives have distinctive colours.

What colour is the 2,4-dinitrophenylhydrazone derivative of propanone?      1

     **(6)**

13. When sodium hydroxide solution was added to 2-bromomethylpropane an $S_N1$ reaction took place producing methylpropan-2-ol and hydrobromic acid.

(a) (i) What is meant by an $S_N1$ reaction?      2

(ii) Draw the structure of the carbocation intermediate formed in this reaction.      1

(b) Chloromethane reacts with sodium ethoxide in an $S_N2$ reaction.

(i) How is sodium ethoxide prepared in the laboratory?      1

(ii) Name the organic product of this $S_N2$ reaction.      1

     **(5)**

*Marks*

**14.** The structure of lactic acid is

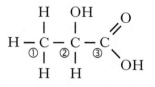

(a) What is the systematic name of lactic acid? 1

(b) Lactic acid contains an asymmetric carbon atom.

Identify, and **explain**, which one of the numbered carbon atoms is asymmetric. 1

(c) Lactic acid can be produced from ethanal by the reaction sequence below.

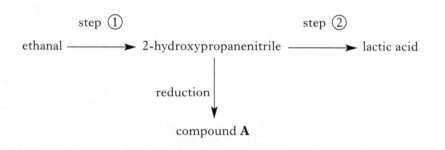

(i) Which reagent could be used in step ① ? 1

(ii) What type of reaction takes place in step ② ? 1

(iii) Draw a structure for compound **A**. 1

**(5)**

*[END OF QUESTION PAPER]*

[BLANK PAGE]

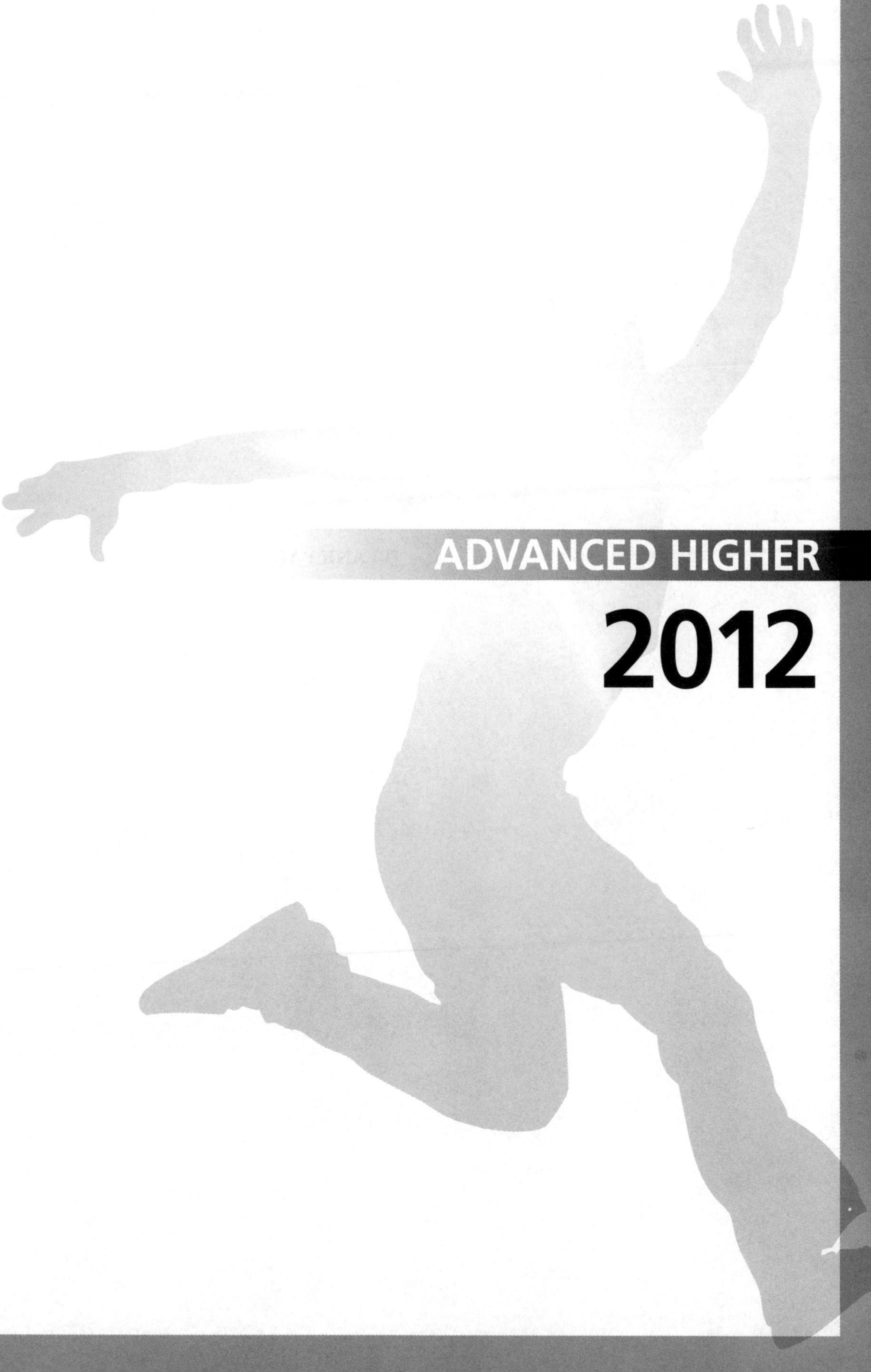

# ADVANCED HIGHER

# 2012

[BLANK PAGE]

# X012/13/02

NATIONAL
QUALIFICATIONS
2012

MONDAY, 14 MAY
1.00 PM – 3.30 PM

CHEMISTRY
ADVANCED HIGHER

Reference may be made to the Chemistry Higher and Advanced Higher Data Booklet.

**SECTION A – 40 marks**

Instructions for completion of **SECTION A** are given on page two.

For this section of the examination you must use an **HB pencil.**

**SECTION B – 60 marks**

All questions should be attempted.

**Answers must be written clearly and legibly in ink.**

## SECTION A

**Read carefully**

1 Check that the answer sheet provided is for **Chemistry Advanced Higher (Section A)**.

2 For this section of the examination you must use an **HB pencil** and, where necessary, an eraser.

3 Check that the answer sheet you have been given has **your name**, **date of birth**, **SCN** (Scottish Candidate Number) and **Centre Name** printed on it.

   Do not change any of these details.

4 If any of this information is wrong, tell the Invigilator immediately.

5 If this information is correct, **print** your name and seat number in the boxes provided.

6 The answer to each question is **either** A, B, C or D. Decide what your answer is, then, using your pencil, put a horizontal line in the space provided (see sample question below).

7 There is **only one correct** answer to each question.

8 Any rough working should be done on the question paper or the rough working sheet, **not** on your answer sheet.

9 At the end of the exam, put the **answer sheet for Section A inside the front cover of your answer book**.

**Sample Question**

To show that the ink in a ball-pen consists of a mixture of dyes, the method of separation would be

   A chromatography

   B fractional distillation

   C fractional crystallisation

   D filtration.

The correct answer is **A**—chromatography. The answer **A** has been clearly marked in **pencil** with a horizontal line (see below).

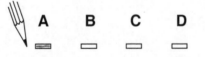

**Changing an answer**

If you decide to change your answer, carefully erase your first answer and using your pencil, fill in the answer you want. The answer below has been changed to **D**.

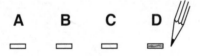

1. Which of the following is **not** a form of electromagnetic radiation?

   A    $\alpha$ radiation

   B    $\gamma$ radiation

   C    UV radiation

   D    X-rays

2. An ion, $X^{3+}$, contains 55 electrons.

   In which block of the Periodic Table would element **X** be found?

   A    s

   B    p

   C    d

   D    f

3. Which of the following statements is **true** about a $Co^{2+}(g)$ ion?

   A    It has 5 unpaired electrons.

   B    It has 8 electrons in s orbitals.

   C    It has 13 electrons in the third shell.

   D    Its electrons with the highest energy are in 3d orbitals.

4. In absorption spectroscopy, as the concentration of an ionic solution decreases, the radiation transmitted

   A    increases in intensity

   B    decreases in intensity

   C    increases in wavelength

   D    decreases in wavelength.

5. Neon gas discharge lamps produce a red glow because electrons in neon atoms are

   A    absorbing radiation from the blue end of the visible spectrum

   B    emitting radiation from the red end of the visible spectrum

   C    emitting radiation from the blue end of the visible spectrum

   D    absorbing radiation from the red end of the visible spectrum.

6. Which of the following molecules has three atoms in a straight line?

   A    $H_2O$

   B    $SF_6$

   C    $CH_4$

   D    $C_2H_3Br$

7. Which of the following ligands is bidentate?

   A    $CN^-$

   B    $NH_3$

   C    $H_2O$

   D    $H_2NCH_2CH_2NH_2$

8. $PCl_5 \rightleftharpoons PCl_3 + Cl_2$

   Adding $PCl_3$ to the above system will

   A    increase the value of the equilibrium constant

   B    decrease the value of the equilibrium constant

   C    increase the concentration of $PCl_5$ and decrease the concentration of $Cl_2$

   D    decrease the concentration of $PCl_5$ and increase the concentration of $Cl_2$.

9. $AgCl(s) \rightarrow Ag^+(aq) + Cl^-(aq)$

   The solubility product ($K_s$) for silver chloride is given by the expression

   $$K_s = [Ag^+(aq)]\,[Cl^-(aq)]$$

   The formula mass of AgCl is 143·4.

   $K_s = 1·80 \times 10^{-10}$ at 25 °C.

   The solubility of silver chloride, in $mol\,l^{-1}$, at 25 °C is

   A    $1·80 \times 10^{-10}$

   B    $3·60 \times 10^{-10}$

   C    $1·34 \times 10^{-5}$

   D    $2·68 \times 10^{-5}$.

**[Turn over**

10. At a particular temperature, 8·0 mole of $NO_2$ was placed in a 1 litre container and the $NO_2$ dissociated by the following reaction:

$$2NO_2(g) \rightleftharpoons 2NO(g) + O_2(g)$$

At equilibrium the concentration of NO(g) is $2·0 \ mol \ l^{-1}$.

The equilibrium constant will have a value of

A    0·11

B    0·22

C    0·33

D    9·00.

11. A buffer solution can **not** be made from

A    $CH_3CH_2COOH$ and $CH_3CH_2COONa$

B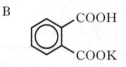

C    $HNO_3$ and $NaNO_3$

D    $NH_3$ and $NH_4Cl$.

12. $5·0 \ cm^3$ of a solution of hydrochloric acid was diluted to exactly $250 \ cm^3$ with water. The pH of this diluted solution was 2·00.

The concentration of the original undiluted solution, in $mol \ l^{-1}$, was

A    $2·0 \times 10^{-2}$

B    $4·0 \times 10^{-2}$

C    $4·0 \times 10^{-1}$

D    $5·0 \times 10^{-1}$.

13. The graph below shows the pH changes when $0·1 \ mol \ l^{-1}$ ammonia solution is added to $50 \ cm^3$ of $0·1 \ mol \ l^{-1}$ hydrochloric acid solution.

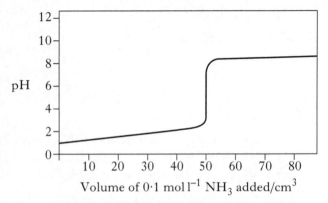

Volume of $0·1 \ mol \ l^{-1} \ NH_3$ added/$cm^3$

Which line in the table shows an indicator which is **not** suitable for use in determining the equivalence point for the above reaction?

|   | Indicator | pH range of indicator |
|---|---|---|
| A | methyl orange | 3·1 – 4·4 |
| B | bromophenol red | 5·2 – 6·8 |
| C | bromothymol blue | 6·0 – 7·6 |
| D | phenolphthalein | 8·3 – 10·0 |

14. $C(s) + O_2(g) \rightarrow CO_2(g) \quad \Delta H° = -396 \ kJ \ mol^{-1}$

$Pb(s) + \frac{1}{2}O_2(g) \rightarrow PbO(s) \quad \Delta H° = -210 \ kJ \ mol^{-1}$

$PbO(s) + CO(g) \rightarrow Pb(s) + CO_2(g) \quad \Delta H° = -74 \ kJ \ mol^{-}$

What is the value of $\Delta H°$, in $kJ \ mol^{-1}$, for the following reaction?

$C(s) + \frac{1}{2}O_2(g) \rightarrow CO(g)$

A    −260

B    −112

C    +112

D    +260

**15.** $50 \, cm^3$ of $1 \, mol \, l^{-1}$ sodium hydroxide is placed in a beaker.

Which of the following graphs shows how the temperature of the solution in the beaker would change as $100 \, cm^3$ of $1 \, mol \, l^{-1}$ hydrochloric acid is gradually added?

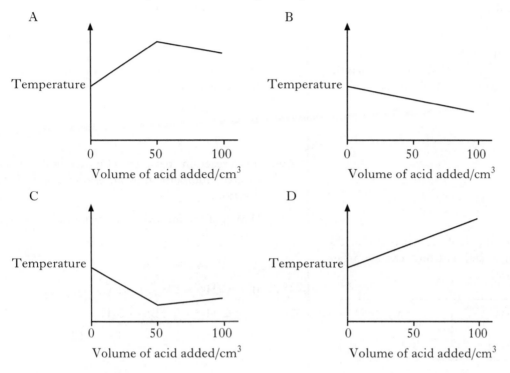

**16.** Which of the following enthalpy changes can **not** be measured directly by experiment?

  A    Enthalpy of formation of methane

  B    Enthalpy of combustion of hydrogen

  C    Enthalpy of formation of carbon dioxide

  D    Enthalpy of combustion of carbon monoxide

**17.**

| Bond | Bond enthalpy/kJ mol$^{-1}$ |
|------|------------------------------|
| H—H | 432 |
| Cl—Cl | 243 |
| H—Cl | 428 |

Using the above data, the standard enthalpy of formation of $HCl(g)$, in $kJ \, mol^{-1}$, is

  A    $-90{\cdot}5$

  B    $-123{\cdot}5$

  C    $-181$

  D    $-247$.

**18.** Which of the following equations represents a step that is **not** involved in the Born Haber cycle for the formation of rubidium iodide?

  A    $I_2(s) \rightarrow I_2(g)$

  B    $I_2(g) \rightarrow 2I(g)$

  C    $I(g) \rightarrow I^+(g) + e^-$

  D    $I(g) + e^- \rightarrow I^-(g)$

**19.** $Cr^+(g) \rightarrow Cr^{3+}(g) + 2e^-$

The energy required for this change per mole of chromium(III) ions is

  A    $2259 \, kJ$

  B    $3000 \, kJ$

  C    $4600 \, kJ$

  D    $5259 \, kJ$.

**[Turn over**

**20.** For any liquid, $\Delta S_{vapourisation} = \dfrac{\Delta H_{vapourisation}}{T_b}$

where $T_b$ = boiling point of that liquid.

For many liquids,

$\Delta S_{vapourisation} = 88\,\text{J K}^{-1}\,\text{mol}^{-1}$.

Assuming that this value is true for water and that its $\Delta H_{vapourisation} = 40 \cdot 6\,\text{kJ mol}^{-1}$, then the boiling point of water is calculated as

A   $0 \cdot 46\,\text{K}$

B   $2 \cdot 17\,\text{K}$

C   $373\,\text{K}$

D   $461\,\text{K}$.

**21.** Which line in the table is correct for the enthalpy change and entropy change when steam condenses?

|   | $\Delta H$ | $\Delta S$ |
|---|---|---|
| A | +ve | +ve |
| B | +ve | −ve |
| C | −ve | −ve |
| D | −ve | +ve |

**22.**

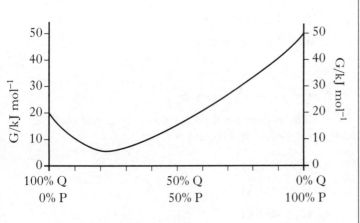

Assuming that liquids P and Q are in their standard states when 100% of either is present, what is the value of $\Delta G°$, in kJ mol$^{-1}$, for the reaction represented by the stoichiometric equation,

$$Q(\ell) \rightarrow P(\ell)?$$

A   −15

B   −30

C   +30

D   +45

**23.** 2-Bromobutane reacts with potassium hydroxide in ethanol to produce two unsaturated products.

The type of reaction involved is

A   addition

B   elimination

C   oxidation

D   substitution.

**24.** The reaction between chlorine and ethane to give chloroethane is an example of a chain reaction.

Which of the following is a propagation step in this reaction?

A   $Cl_2 \rightarrow Cl\bullet + Cl\bullet$

B   $C_2H_5\bullet + Cl\bullet \rightarrow C_2H_5Cl$

C   $C_2H_5\bullet + C_2H_5\bullet \rightarrow C_4H_{10}$

D   $C_2H_5\bullet + Cl_2 \rightarrow C_2H_5Cl + Cl\bullet$

**25.** Which of the following molecules is likely to produce the most stable carbocation intermediate in a substitution reaction?

A   $CH_3CH_2I$

B   $(CH_3)_3CCl$

C   $CH_3CH_2Cl$

D   $CH_3CHICH_2CH_3$

**26.** Which of the following compounds will have the highest boiling point?

A   H–C–C–C–C with H H H on top, H H H below, and $\begin{smallmatrix}O\\OH\end{smallmatrix}$

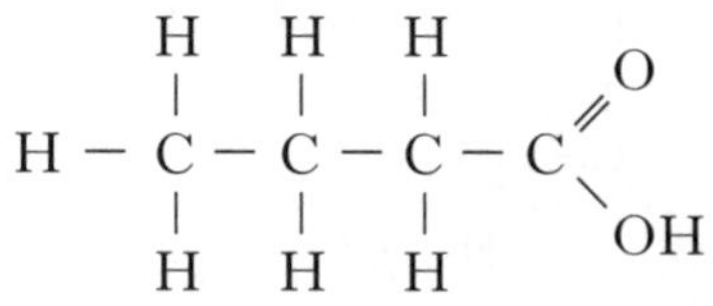

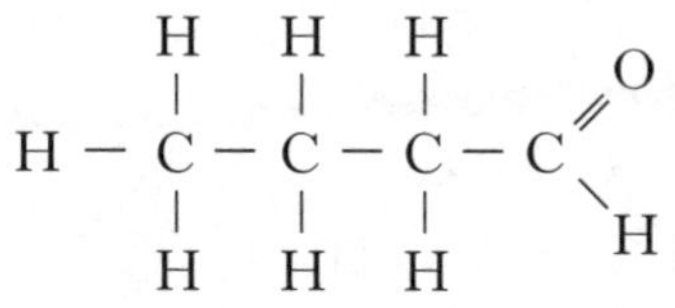

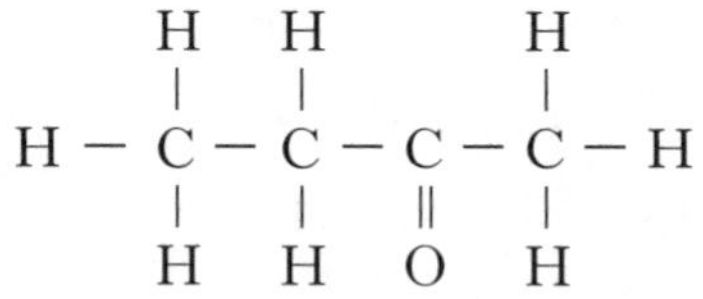

B   H–C–C–C–C with H H H on top, H H H below, and $\begin{smallmatrix}O\\H\end{smallmatrix}$

C   H–C–C–C–C–H with H H H on top, H H O H below

D   H–C–C–C–C–OH with H H H H on top, H H H H below

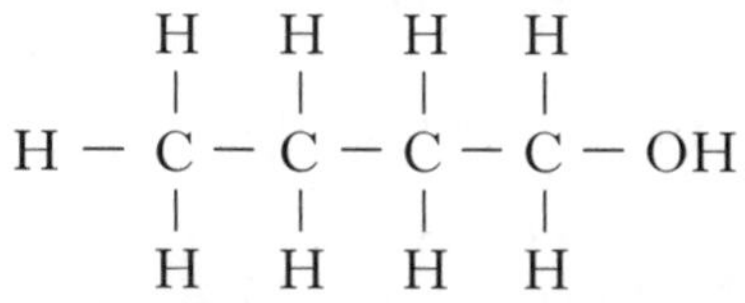

**27.** Which of the following is an **essential** property of a solvent to be used for recrystallisation purposes?

A   Insoluble in water

B   A low boiling point

C   Ability to dissolve more solute when hot than when cold

D   Ability to dissolve more solute when cold than when hot

**28.** Which of the following is correct for the reaction of propene with hydrogen bromide?

A   1-Bromopropane is the only product.

B   1-Bromopropane is the major product.

C   2-Bromopropane is the only product.

D   2-Bromopropane is the major product.

**29.** Hybrid orbitals can be formed by the mixing of s and p orbitals.

Which of the following hybrid orbitals are most likely to be involved in the bonding in ethyne?

A   sp

B   $sp^2$

C   $sp^3$

D   $s^2p$

**30.** Carbon dioxide has the following structure.

$$O = C = O$$

Which line in the table shows the correct numbers of $\sigma$ and $\pi$ bonds in a molecule of carbon dioxide?

|   | Number of $\sigma$ bonds | Number of $\pi$ bonds |
|---|---|---|
| A | 0 | 2 |
| B | 2 | 2 |
| C | 4 | 0 |
| D | 0 | 4 |

**31.** P   (benzene ring)–Cl

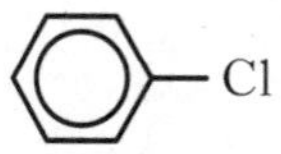

Q   $CH_2=CHCl$

R   $CH_2=CHCH_2Cl$

Which of the above molecules is/are planar?

A   **P** only

B   **P** and **Q** only

C   **Q** and **R** only

D   **P**, **Q** and **R**

**[Turn over**

**32.** Which of the following can be distinguished by making 2,4-dinitrophenylhydrazone derivatives?

    A    Ethanal and propanal

    B    Propan-1-ol and propan-2-ol

    C    Ethanoic acid and benzoic acid

    D    Methoxyethane and ethoxyethane

**33.** Which of the following could be the molecular formula for a ketone?

    A    $C_3H_8O$

    B    $C_3H_6O$

    C    $C_2H_4O$

    D    $CH_2O$

**34.** Which of the following compounds would dissolve in water to give an alkaline solution?

    A    $CH_3CH_2CN$

    B    $CH_3CH_2CHO$

    C    $CH_3CH_2CH_2OH$

    D    $CH_3CH_2CH_2NH_2$

**35.**

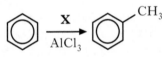

Which of the following compounds could be **X**?

    A    $CH_4$

    B    $CH_3Cl$

    C    $CH_2Cl_2$

    D    $CH_3OH$

**36.** Which of the following has a geometric isomer?

    A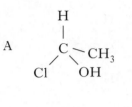

    B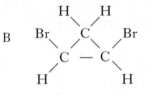

    C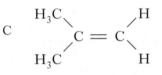

    D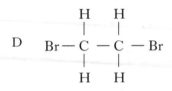

**37.** Combustion analysis of hydrocarbon **X** showed that it contained 82·7% carbon and 17·3% hydrogen.

The molecular formula for **X** could be

    A    $CH_3$

    B    $C_2H_6$

    C    $C_2H_5$

    D    $C_4H_{10}$.

**38.** The number of waves per centimetre is known as the

    A    wavenumber

    B    wavelength

    C    frequency

    D    intensity.

**39.** Which of the following analytical techniques depends on the vibrations within molecules?

A    Colorimetry

B    Mass spectroscopy

C    Proton nmr spectroscopy

D    Infra-red absorption spectroscopy

**40.**

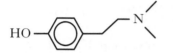

The active structural fragment of several pain-killing molecules is shown.

What term best describes this structural fragment?

A    Agonist

B    Receptor

C    Antagonist

D    Pharmacophore

*[END OF SECTION A]*

**Candidates are reminded that the answer sheet for Section A MUST be placed INSIDE the front cover of your answer book.**

**[Turn over for SECTION B on *Page ten***

## SECTION B

*Marks*

**60 marks are available in this section of the paper.**

**All answers must be written clearly and legibly in ink.**

1. Semiconductors are used in a wide variety of applications.

   (a) In Blu-ray DVD players, light of wavelength 405 nm is produced from a gallium(III) nitride laser.

       (i) Calculate the energy, in kJ mol$^{-1}$, corresponding to this wavelength.     **2**

       (ii) Write the electronic configuration of gallium(III) in terms of s, p and d orbitals.     **1**

   (b) The electrical conductivity of the semiconductor gallium arsenide increases on exposure to light.

   What name is given to this phenomenon?     **1**

   (c) Doped silicon is also used as a semiconductor.

   What is the main current carrier in silicon doped with boron?     **1**

       **(5)**

2. The nitrate ion, $NO_3^-$, can be converted into either nitrous acid, $HNO_2$ or nitrogen monoxide, NO.

   The oxidation state of nitrogen in NO is +2.

   (a) Calculate the oxidation state of nitrogen in

       (i) $NO_3^-$

       (ii) $HNO_2$.     **1**

   (b) Write a balanced ion-electron equation for the reduction of nitrous acid into the compound $H_2N_2O_2$.     **1**

   (c) Nitrogen is also present in the cyanide ion, $CN^-$.

   Name the complex ion $[Cu(CN)_2]^-$.     **1**

       **(3)**

*Marks*

3. Two common crystal lattice structures adopted by ionic compounds can be described as simple cubic and face-centred cubic.

   (a) What determines the type of structure adopted by a particular ionic compound?    **1**

   (b) Sodium chloride has a face-centred cubic structure which has 6:6 coordination.

   Explain what 6:6 coordination means.    **1**

   (c) Caesium chloride has a simple cubic structure which has 8:8 coordination.

   Which potassium halide is most likely to have 8:8 coordination?    **1**

   (d) Many ionic compounds are soluble in water.

      (i) Which two factors determine whether the enthalpy of solution is exothermic or endothermic?    **1**

      (ii) The enthalpy of solution of sodium chloride is $0 \, kJ \, mol^{-1}$.

         Suggest what makes the dissolving of sodium chloride in water a feasible process.    **1**

   **(5)**

4. $BH_3$ in the gas phase is very reactive. It readily combines with the compound tetrahydrofuran, $C_4H_8O$, to make a more stable compound.

$$BH_3 + C_4H_8O \rightarrow C_4H_8OBH_3$$

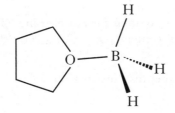

   (a) What is the shape of a $BH_3$ molecule?    **1**

   (b) In the more stable compound a dative covalent bond exists between the boron and oxygen.

   How does this dative covalent bond form?    **1**

   (c) To which class of organic compound does tetrahydrofuran belong?    **1**

   **(3)**

**[Turn over**

*Marks*

5. As part of an investigation a student was analysing the metallic content of a key known to be composed of a copper/nickel alloy.

   The key was dissolved in nitric acid and the resulting solution diluted to $1000 \, cm^3$ in a standard flask using tap water. Three $25 \cdot 0 \, cm^3$ samples of the nitrate solution were pipetted into separate conical flasks and approximately $10 \, g$ of solid potassium iodide were added. Iodine was produced as shown in the equation.

   $$2Cu^{2+}(aq) + 4I^-(aq) \rightarrow 2CuI(s) + I_2(aq)$$

   The liberated iodine was titrated against standardised $0 \cdot 102 \, mol \, l^{-1}$ sodium thiosulphate solution. Starch indicator was added near the end point of the titration.

   $$I_2(aq) + 2S_2O_3^{2-}(aq) \rightarrow 2I^-(aq) + S_4O_6^{2-}(aq)$$

   The results, for the volume of thiosulphate used, are given in the table.

   |  | Titration 1 | Titration 2 | Titration 3 |
   |---|---|---|---|
   | Final volume/$cm^3$ | 16·30 | 31·50 | 46·80 |
   | Initial volume/$cm^3$ | 0·30 | 16·30 | 31·50 |
   | Volume added/$cm^3$ | 16·00 | 15·20 | 15·30 |

   (a) From the results calculate the mass of copper in the key. 3

   (b) Suggest how the accuracy of the analysis could be improved. 1

   (c) The student then tried to analyse the original nitrate solution for nickel using EDTA as in a PPA experiment. The value obtained for the nickel content was much greater than the true value.

   Give the main reason why the value obtained was higher than the true value. 1

   **(5)**

*Marks*

**6.** A student was trying to determine the partition coefficient of propanedioic acid between the two solvents, hexane and water.

$$\text{propanedioic acid}_{(water)} \rightleftharpoons \text{propanedioic acid}_{(hexane)}$$

The following series of steps were carried out.

   Step A.  25 cm³ water and 25 cm³ hexane were pipetted into apparatus X.

   Step B.  A measured mass of propanedioic acid was added to the solvents in apparatus X.

   Step C.  The mixture was shaken for approximately 2 minutes and allowed to settle.

These steps were repeated with different masses of propanedioic acid.

(a) Name apparatus X.                                                                                                    1

(b) A series of titrations were carried out which enabled the student to work out the equilibrium concentrations of propanedioic acid in the two solvents. The values obtained are given in the table below.

| Mass of propanedioic acid used/g | Concentration of propanedioic acid in water/mol l⁻¹ | Concentration of propanedioic acid in hexane/mol l⁻¹ |
|:---:|:---:|:---:|
| 0·31 | 0·24 | 0·031 |
| 0·44 | 0·30 | 0·038 |
| 0·61 | 0·37 | 0·048 |

Use these results to calculate a value for the partition coefficient.                        1

(c) The student repeated the experiment several weeks later using the same chemicals. The values obtained are given in the table below.

| Mass of propanedioic acid used/g | Concentration of propanedioic acid in water/mol l⁻¹ | Concentration of propanedioic acid in hexane/mol l⁻¹ |
|:---:|:---:|:---:|
| 0·93 | 0·57 | 0·083 |

Give the reason why this experiment produces a different value for the partition coefficient compared to the value calculated earlier.                                                            1

(d) Why would no partition take place if ethanol had been used instead of hexane?        1

                                                                                                                                (4)

**[Turn over**

*Marks*

7. Balsamic vinegar is a dark brown liquid containing ethanoic acid. The pH of a sample of balsamic vinegar was 2·5.

   (a) Calculate the concentration of ethanoic acid in the sample of balsamic vinegar.    **2**

   (b) A student chose to use a pH meter rather than use an indicator for the titration of balsamic vinegar with sodium hydroxide.

   Apart from being more accurate, suggest why the student chose to use a pH meter rather than an indicator for this particular titration.    **1**

   (c) Write the formula for the conjugate base of ethanoic acid.    **1**

   **(4)**

8. Part of an Ellingham diagram is shown below.

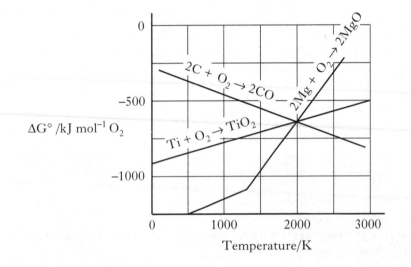

   (a) Using the Ellingham diagram give the temperature **range** over which magnesium will reduce titanium dioxide to titanium.    **1**

   (b) Suggest why the line labelled $2C + O_2 \rightarrow 2CO$ slopes downward.    **1**

   (c) Suggest why the gradient of the line labelled $2Mg + O_2 \rightarrow 2MgO$ changes at approximately 1360 K.    **1**

   **(3)**

9. Silver oxide cells are used in hearing aids. Zinc is the negative electrode and silver(I) oxide is the positive electrode. The overall cell reaction is represented by the equation

   $$Zn(s) + Ag_2O(s) \rightarrow ZnO(s) + 2Ag(s)$$

   The free energy change for the cell is −279·8 kJ per mole of zinc.

   Calculate the voltage produced by the cell.    **(3)**

*Marks*

**10.** The graphs show how the concentrations of reactants A and B change with time for the reaction

$$A + B \rightarrow C$$

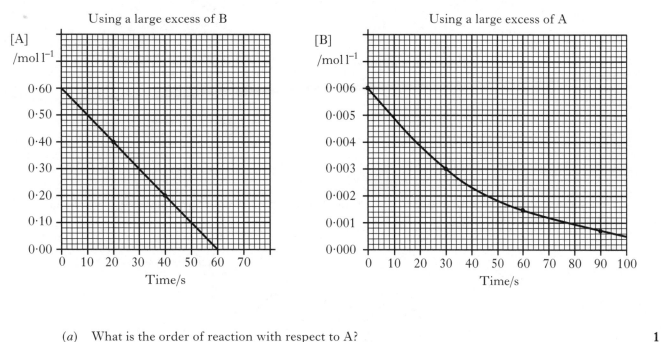

(a) What is the order of reaction with respect to A?    1

(b) What is the order of reaction with respect to B?    1

(c) What are the units of the rate constant in this reaction?    1

**(3)**

**11.** Both lithium aluminium hydride, $LiAlH_4$, and phosphorus pentachloride, $PCl_5$, react vigorously with water producing different gases.

(a) Name the gas produced when water reacts with

(i) lithium aluminium hydride    1

(ii) phosphorus pentachloride.    1

(b) Phosphorus pentachloride will also react with any compound containing a hydroxyl group. A chlorine atom replaces the hydroxyl group. For example,

$$C_6H_5COOH \xrightarrow{PCl_5} C_6H_5COCl \quad \text{or} \quad CH_3COOH \xrightarrow{PCl_5} CH_3COCl$$

(i) What type of organic compound is produced in these reactions?    1

(ii) Draw a structural formula for the ester formed when $C_6H_5COCl$ reacts with propan-2-ol.    1

(iii) What is the advantage of using $C_6H_5COCl$ instead of benzoic acid in this esterification reaction?    1

**(5)**

*Marks*

12.  Skeletal structural formulae are used to show structures of molecules more simply than full structural formulae.

For example, pent-1-ene can be represented as

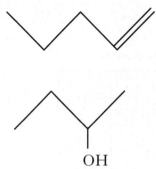

and butan-2-ol as

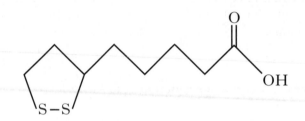

OH

Lipoic acid has recently been used as a food supplement. The skeletal structural formula of lipoic acid is shown below.

O

OH

S—S

(a)  Write the molecular formula of lipoic acid.                                                                1

(b)  (i)  Lipoic acid is optically active. Copy the skeletal structural formula of lipoic acid and circle the carbon atom responsible for the optical activity of lipoic acid.          1

(ii)  Why does this carbon atom make lipoic acid optically active?                          1

(3)

*Marks*

**13.** In a PPA, benzoic acid is prepared from ethyl benzoate by refluxing with sodium hydroxide solution.

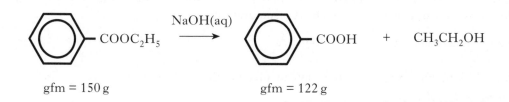

gfm = 150 g                     gfm = 122 g

(*a*)   Why is the mixture refluxed rather than heated in an open beaker?     1

(*b*)   Name the type of reaction that is involved between ethyl benzoate and sodium hydroxide solution.     1

(*c*)   What does the procedure suggest should be added to the flask along with ethyl benzoate and sodium hydroxide solution?     1

(*d*)   What change in appearance of the contents of the flask indicates that the reaction is complete?     1

(*e*)   A yield of 73·2% of benzoic acid was obtained from 5·64 g of ethyl benzoate.

Calculate the mass of benzoic acid produced.     2

**(6)**

**14.** (*a*)   Benzene reacts with a "nitrating mixture" to produce nitrobenzene.

(i)   Name the type of chemical reaction that takes place in the nitration of benzene.     1

(ii)   Nitrobenzene is reduced by reaction with a mixture of tin and concentrated hydrochloric acid to form an organic base.

Identify this organic base.     1

(*b*)   Benzene also reacts with sulphur trioxide dissolved in concentrated sulphuric acid to produce benzenesulphonic acid, $C_6H_5SO_3H$.

(i)   Draw a structural formula for benzenesulphonic acid.     1

(ii)   Draw a Lewis electron dot diagram for sulphur trioxide.     1

**(4)**

**[Turn over for Question 15 on *Page eighteen***

*Marks*

**15.** Chloroalkane **A** has molecular formula $C_4H_9Cl$. When **A** is heated with NaOH(aq), it undergoes an $S_N2$ reaction to form alcohol **B**.

Alcohol **B** can be oxidised by acidified potassium dichromate solution and it can also be dehydrated to produce a mixture of two alkenes which are structural isomers.

(*a*) Draw a structural formula for compound **A**.     1

(*b*) Draw the structure of the transition state involved in this $S_N2$ reaction.     1

(*c*) The simplified proton nmr spectrum of one of the alkenes is shown.

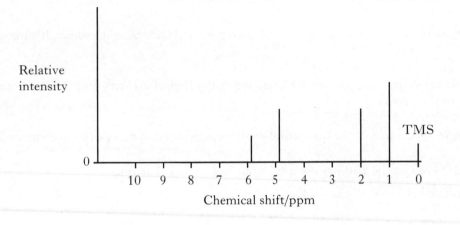

Sketch the proton nmr spectrum of the other alkene.     2

    **(4)**

[*END OF QUESTION PAPER*]

# SQA ADVANCED HIGHER
# CHEMISTRY 2008–2012

## CHEMISTRY ADVANCED HIGHER
## 2008

### SECTION A

| | | | |
|---|---|---|---|
| 1. D | 11. B | 21. D | 31. D |
| 2. D | 12. D | 22. B | 32. B |
| 3. B | 13. D | 23. A | 33. D |
| 4. C | 14. C | 24. A | 34. D |
| 5. B | 15. B | 25. C | 35. A |
| 6. A | 16. C | 26. D | 36. B |
| 7. D | 17. A | 27. B | 37. A |
| 8. C | 18. A | 28. A | 38. C |
| 9. A | 19. B | 29. C | 39. D |
| 10. C | 20. A | 30. C | 40. C |

### SECTION B

1. (a) Octahedral/square bipyramidal **or**
   5 bonded pairs and 1 lone pair

   (b) $NF_3$ has 4 electron pairs / $BF_3$ has 3 electron pairs
   $NF_3$ has tetrahedral arrangement of electron pairs / $BF_3$ has trigonal planar arrangement of electron pairs
   $NF_3$ has an extra electron pair / N has an extra electron pair

2. (a) Green

   (b) (i) +3 or 3+ or 3 or III or three
   (ii) Tetraamminedichlorocobalt(III)
   (iii) $1s^2 2s^2 2p^6 3s^2 3p^6 3d^6$

3. (a) (Excited) electrons emitting energy as they fall back down to lower energy levels

   (b) Wavelength of emitted light is outwith the visible part of the spectrum
   Emission is in U.V. or I.R.
   Flame not hot enough

   (c) $178.45$ kJ mol$^{-1}$

4. (a) (i) It coordinates through 2 sites to the platinum
   Forms 2 bonds to the Pt
   Pt is attached to 2 parts of the DNA
   Forms 2 dative bonds to the Pt
   (ii) Lone pairs of electrons
   Non bonded pairs of electrons
   Unbonded pairs of electrons

   (b)
   Cl＼　　＼NH₃
   　　　Pt
   H₃N＼　　＼Cl

5. (a) (i) $-478$ kJ mol$^{-1}$
   (ii) $217$ J K$^{-1}$ mol$^{-1}$
   (iii) $-542.7$ kJ mol$^{-1}$ **or** $-543$ kJ mol$^{-1}$

   (b) $-570$ kJ mol$^{-1}$

6. (a) $1.025 \times 10^{-3}$ **or** $1.03 \times 10^{-3}$

   (b) $1.025$ mol l$^{-1}$

7. (a) Step **X** $= 77$ kJ **or** 77 kJ mol$^{-1}$
   Step **Y** $= 382$ kJ **or** 382 kJ mol$^{-1}$

   (b) Lattice enthalpy
   Lattice formation enthalpy

   (c) $-535.5$ kJ mol$^{-1}$

8. (a) Starch solution **or** starch **or** iotec

   (b) (i) $2.35 \times 10^{-2}$ mol l$^{-1}$
   (ii) $2.625 \times 10^{-2}$ mol l$^{-1}$

   (c) $0.894$

   (d) (i) It decreases
   (ii) It would stay the same/No change/No effect

9. (a) $K_a = \dfrac{[H_3O^+(aq)] \times [F^-(aq)]}{[HF(aq)]}$

   (b) From the graph, $pK_a = 3.8$
   therefore $K_a = 1.58 \times 10^{-4}$

   (c) Sodium fluoride
   NaF **or** Na$^+$F$^-$ **or** F$^-$

   (d) Cresol red / alizarin red

10. (a) Elimination

   (b) (i) $sp^2$ hybridisation is a mixing of one s orbital and two p orbitals
   (ii) Sigma bonds – end on overlap of (atomic) orbitals
   Pi bonds – sideways overlap of (atomic) orbitals

   (c)
```
     H    H    H    H
     |    |    |    |
  H—C——C——C——C—H
     |    |    |    |
     H    H   OH    H
```

   (d) (i) 2nd order
   (ii) $1.32 \times 10^{-4}$ l mol$^{-1}$ s$^{-1}$
   (iii)
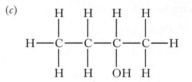

11. (a) (i) Sodium metal
   any reactive metal
   any Group 1 metal
   Ba
   (ii) (Hot) copper(II) oxide
   acidified dichromate
   acidified permanganate
   copper oxide
   acidified chromate

   (b) Ethoxypropane

(c)

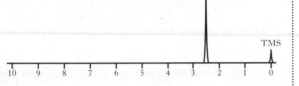

12. (a) Condensation

   **or** addition + elimination

(b) (i) By recrystallisation/crystallisation
   (ii) Measure melting point of derivative and compare with literature values/expected value/Data Book value/known value

(c) (i) Fehling's solution and blue to orange/brown with the isomer (propanal)
   Benedict's becomes orange/red with the isomer
   Tollens' reagent gives silver mirror with the isomer
   Acidified dichromate changes from orange to green with the isomer
   Acidified permanganate changes from purple to colourless with the isomer
   (Hot) copper(II) oxide changes from black to brown with the isomer

(ii)

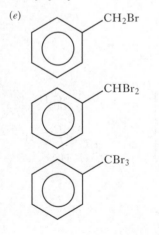

(iii) A = $CH_3^+$
   B = $CH_3CO^+$ **or** $C_2H_3O^+$

13. (a) Electrophilic substitution

(b) $Br_2$ and $FeBr_3/FeCl_3/AlBr_3/AlCl_3$

(c) Sulphuric acid and nitric acid
   $H_2SO_4$ and $HNO_3$
   Concentrated/fuming $H_2SO_4 + HNO_3$

(d) $C_6H_6SO_3$

(e)

## CHEMISTRY ADVANCED HIGHER 2009

### SECTION A

| | | | |
|---|---|---|---|
| 1. | D | 21. | B |
| 2. | D | 22. | C |
| 3. | A | 23. | A |
| 4. | C | 24. | C |
| 5. | A | 25. | B |
| 6. | C | 26. | D |
| 7. | D | 27. | D |
| 8. | B | 28. | A |
| 9. | D | 29. | C |
| 10. | C | 30. | C |
| 11 | D | 31. | B |
| 12. | D | 32. | C |
| 13. | A | 33. | C |
| 14. | A | 34. | D |
| 15. | B | 35. | B |
| 16. | C | 36. | A |
| 17. | A | 37. | B |
| 18. | A | 38. | B |
| 19. | D | 39. | C |
| 20. | B | 40. | C |

### SECTION B

1. (a) $1s^2 2s^2 2p^6 3s^2 3p^6$

   (b) (i) E= 78.3 (nm)

   (ii) $Ar(g) \rightarrow Ar^+(g) + e^-$

2. (a) 138 J K$^{-1}$ mol$^{-1}$

   (b) 96 kJ mol$^{-1}$

   (c) = 696 K

3. (a) $Mg^{2+}(aq)$

   (b) Lattice (breaking) enthalpy

   (c) −728 kJ

   (d) −322 kJ mol$^{-1}$

4. (a) Bond breaking H-H + $^1/_2$ (O=O)

      = 432 + 248·5 = 680·5
      Bond making 2 O-H = −916
      $\Delta H$ = (680·5 −916) = −235·5 (kJ mol$^{-1}$)

   (b) The above reaction has formed $H_2O(g)$ and more energy will be given out as it changes to $H_2O(l)$
      **or**
      Enthalpy of combustion forms $H_2O(l)$ at standard conditions but the above reaction has formed $H_2O(g)$

5. (a) 1 mol l$^{-1}$ $H^+$ ions, 25°C (298K) and 1 atmosphere pressure

   (b) $2IO_3^- + 12H^+ + 10e^- \rightarrow I_2 + 6H_2O$

   (c) −574·2 kJ mol$^{-1}$

6. (a) (i) $HCOO^-$ **or** methanoate ion

   (ii) $K_a = \dfrac{[HCOO^-][H_3O^+]}{[HCOOH]}$

   (b) (i) 0·0783 mol l$^{-1}$

   (ii) 2·43

**7.** (*a*) Rate = $k[CH_3COCH_3][H^+]$

(*b*) The $H^+$ is present at the **start** and the **end** of the reaction

(*c*) (i) To quench/stop the reaction.
To neutralise the acid.

(ii) Starch solution **and** blue/black to colourless

**8.** (*a*) EDTA

(*b*) nickel(II) ions are green **or** green/blue **or** coloured
$Ni^{2+}(aq)$ absorb visible light

(*c*) (i) it has lone pairs of electrons/non-bonding pairs of electrons

(ii) 4

(iii) gravimetric

(iv) to prevent the complex from absorbing moisture
**or**
to allow the complex to cool in a dry atmosphere

**9.** (*a*)

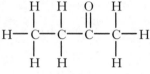

**or**

$CH_3CH_2COCH_3$

(*b*) (nucleophilic) substitution

(*c*) Find its melting point and check with literature values

(*d*)

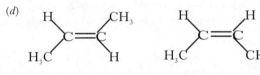

*trans*-but-2-ene            *cis*-but-2-ene

**10.** (*a*) $H_2SO_4$ and $HNO_3$
**or** $H_2SO_4$ and $NaNO_3$

(*b*) reduction

(*c*) Ethanoic acid/$CH_3COOH$
**or** ethanoyl chloride/$CH_3COCl$
**or** ethanoic anhydride/$(CH_3CO)_2O$

**11.** (*a*) $C_4H_8O_2$

(*b*) (i) carbonyl **or** C=O

(ii) ester

(*c*) $[CH_3CH_2CO]^+$
$C_2H_5CO^+$

(*d*) methyl propanoate

**12.** (*a*) +5 and +7

(*b*) trigonal bipyramidal

(*c*) $sp^3d$ **or** $sp^2d^2$ **or** $spd^3$

(*d*) Cl atom too small to accommodate 7 F atoms around it

**13.** (*a*)

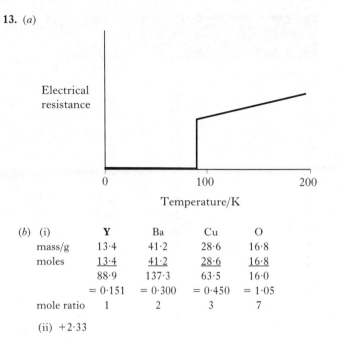

(*b*) (i)

| | Y | Ba | Cu | O |
|---|---|---|---|---|
| mass/g | 13·4 | 41·2 | 28·6 | 16·8 |
| moles | $\frac{13·4}{88·9}$ | $\frac{41·2}{137·3}$ | $\frac{28·6}{63·5}$ | $\frac{16·8}{16·0}$ |
| | = 0·151 | = 0·300 | = 0·450 | = 1·05 |
| mole ratio | 1 | 2 | 3 | 7 |

(ii) +2·33

(iii) $Y_2Ba_4Cu_6O_{13}$ **or** $YBa_2Cu_3O_6$

## CHEMISTRY ADVANCED HIGHER 2010

### SECTION A

| | | | |
|---|---|---|---|
| 1. | D | 21. | D |
| 2. | C | 22. | C |
| 3. | D | 23. | C |
| 4. | B | 24. | A |
| 5. | B | 25. | C |
| 6. | D | 26. | D |
| 7. | A | 27. | C |
| 8. | B | 28. | B |
| 9. | D | 29. | A |
| 10. | C | 30. | B |
| 11 | C | 31. | D |
| 12. | A | 32. | B |
| 13. | B | 33. | A |
| 14. | A | 34. | B |
| 15. | D | 35. | D |
| 16. | A | 36. | A |
| 17. | D | 37. | C |
| 18. | B | 38. | B |
| 19. | C | 39. | A |
| 20. | D | 40. | C |

### SECTION B

1. (a) $E = \dfrac{Lhc}{\lambda}$ **or** $\dfrac{Lhc}{1000\lambda}$

   $E = \dfrac{6\cdot02 \times 10^{23} \times 6\cdot63 \times 10^{-34} \times 3\cdot00 \times 10^{8}}{160 \times 10^{-9}}$

   $E = 748$ **(kJ mol$^{-1}$) or 748361J**

   (b) (i) 5

   (ii) trigonal bipyramidal

2. (a) **+3 or III or 3**

   (b) tetraaquadichlorochromium(III)

   (c)

   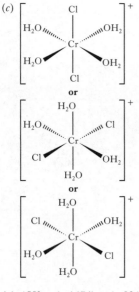

3. (a) $\Delta H° = (-1676) - (-824) =$ **−852 kJ**

   (b) $\Delta S° = [2(27) + 51] - [2(28) + 87] =$ **−38 JK$^{-1}$**

   (c) $\Delta G° = \Delta H° - T\Delta S° = (-852) - 298(-38/1000)$
   $= -852 + 11\cdot32 =$ **−841kJ**

4. (a) (i) Brown fumes, fizzing, solution turning yellow, $NO_2$ forming

   (ii) Oxidising agent

   (iii) Max absorbance of permanganate/Green is the complementary colour of purple

(b) moles of Mn $= 1\cdot4 \times 10^{-4} \times 0\cdot1$
    $= \mathbf{1\cdot4 \times 10^{-5}}$

   mass of Mn $= 1\cdot4 \times 10^{-5} \times 54\cdot9$
   $= \mathbf{7\cdot686 \times 10^{-4}\,g}$

   % Mm $= (7\cdot686 \times 10^{-4}/0\cdot19) \times 100$
   $= \mathbf{0\cdot40\%}$

5. (a)

   $$\left[\begin{array}{c} O \\ \| \\ N \\ \diagup \;\;\; \diagdown \\ O \quad\quad O \end{array}\right]^{-} \quad \text{or} \quad \left[\begin{array}{c} O \\ \| \\ N \\ \diagup \;\;\; \diagdown \\ O \quad\quad O \end{array}\right]^{-}$$

   (b) b = 0
       c = −1 and d = −1

6. (a) (i) colourless to pink/colourless to purple

   (ii) $(16\cdot5/1000) \times 0\cdot02 \times 5/2 =$ **0·000825moles**
   $(8\cdot25 \times 10^{-4})$

   (iii) $0\cdot000825 \times (1000/20) \times 88 =$ **3·63g**

   (iv) $4\cdot49 - 3\cdot63 - 0\cdot06 =$ **0.8g**

   (b)

   | K | H | $C_2O_4$ |
   |---|---|---|
   | 0·8/39 | 0·06/1 | 3·63/88 |
   | 0·020 | 0·060 | 0·041 |

   **X = 1   Y = 3   Z = 2**

7. (a)
   $CH_3CH_2OH + CH_3COOH \rightleftharpoons CH_3COOCH_2CH_3 + H_2O$

   (b) (i) No equilibrium in open system/System will not reach equilibrium

   (ii) At equilibrium:
   moles of water and ester $= 0\cdot70 - 0\cdot24 =$ **0·46**
   moles of ethanoic acid $= 0\cdot24$
   moles of ethanol $= 0\cdot68 - 0\cdot46 =$ **0·22**
   $K = [0\cdot46][0\cdot46]/[0\cdot24][0\cdot22] =$ **4·0**

8. (a)

   (b) $pKa = -logKa = -log\,1\cdot4 \times 10^{-5} = 4\cdot85$
   $pH = \frac{1}{2}\,pKa - \frac{1}{2}\,logc$
   Substitute values   $3\cdot77 = 2\cdot43 - \frac{1}{2}\,logc$
   **c = 0·0020mol l$^{-1}$**

9. (a) Step two
   **or**
   $NO_2 \;+\; F \;\rightarrow\; NO_2F$

   (b) $2NO_2 \;+\; F_2 \;\rightarrow\; 2NO_2F$
   **or**
   $NO_2 \;+\; \frac{1}{2}F_2 \;\rightarrow\; NO_2F$

   (c) 2$^{nd}$ order **or** 2

   (d) $k =$ Initial rate/$[NO_2][F_2] =$ **40 $\ell$ mol $^{-1}$s$^{-1}$**

10. (a) to give a higher yield
    **or**
    to reduce side reactions
    **or**
    to prevent charring

    (b) sodium chloride (solution)/brine/salt water

    (c) to dry the cyclohexene/dry the organic layer/drying agent/absorbs water/removes water

(d) Theoretical mass
of cyclohexene $= \dfrac{82 \times 22 \cdot 56}{100} = 18.5 \text{ g}$

% yield $= \dfrac{6 \cdot 52 \times 100}{18 \cdot 5} = 35 \%$

**or**

Moles cyclohexanol $= 22 \cdot 56/100 = 0 \cdot 2256$ mol
Moles cyclohexane $= 6 \cdot 52/82 = 0 \cdot 0795$ mol

% yield $= \dfrac{0 \cdot 0795}{0 \cdot 2256} \times 100 = 35 \cdot 2 \%$  **or**  35%

**11.** (a) because but-2-ene has two different groups attached to each of the carbon atoms of the double bond
**or**
because in but-1-ene one of the carbon atoms of the double bond has identical groups (H) attached

(b)

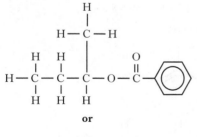

(c) aqueous potassium (or sodium) hydroxide KOH(aq) or NaOH(aq) or LiOH(aq)
**or**
potassium (or sodium) hydroxide solution
**or**
aqueous alkali
**or**
alkali solution
**or**
water/$H_2O$

(d) aluminium chloride or $A\ell C\ell_3$
**or**
iron(III) chloride or $FeC\ell_3$
**or**
iron(III) bromide or $FeBr_3$
**or**
aluminium bromide or $A\ell Br_3$

(e)

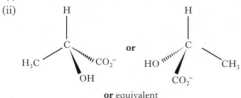

**or**

$CH_3CH_2\overset{\displaystyle CH_3}{\overset{|}{CH}} - O - \overset{\displaystyle O}{\overset{||}{C}} - \bigcirc$

**12.** (a) ethanal

(b) cyanohydrin

(c) hydrolysis/acid hydrolysis

(d) (i) van der Waals' forces
(ii)

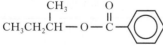

**or** equivalent

Must be tetrahedral but dots and wedges can be replaced by solid lines

(iii) while one group would be able to bind to the appropriate region, the other two would not
**or**
the 3 'functional' groups fail to match the binding regions of the active site
**or**
only 1 group or 2 groups could bind (or match) the binding regions
**or**
The groups on the lactate ion no longer match the binding regions on the active site of the enzyme
**or**
The lactate ion no longer complements the binding region (of the active site)
**or**
The groups now fail to match the binding region (of the active site)

**13.** (a) Alcohols or alkanols
**and**
ethers

(b) (i)

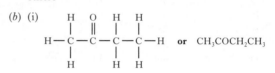

(ii) butan-2-ol

## CHEMISTRY ADVANCED HIGHER 2011

### SECTION A

| | | | |
|---|---|---|---|
| 1. | A | 21. | B |
| 2. | B | 22. | A |
| 3. | A | 23. | D |
| 4. | C | 24. | B |
| 5. | C | 25. | D |
| 6. | C | 26. | C |
| 7. | A | 27. | A |
| 8. | D | 28. | C |
| 9. | C | 29. | A |
| 10. | D | 30. | C |
| 11. | B | 31. | A |
| 12. | C | 32. | B |
| 13. | A | 33. | D |
| 14. | C | 34. | B |
| 15. | C | 35. | D |
| 16. | D | 36. | B |
| 17. | B | 37. | C |
| 18. | B | 38. | A |
| 19. | A | 39. | D |
| 20. | D | 40. | D |

### SECTION B

**1.** (a) Superconductivity
Superconducting
Superconductor
Superconductance

(b) Liquid nitrogen/$N_2$

**2.** (a) The line at $4.6 \times 10^{14}$ Hz

(b) (i) $H(g) \rightarrow H^+(g) + e^-$
$H(g) - e^- \rightarrow H^+(g)$

(ii) $E = \dfrac{Lhc}{\lambda}$ **or** $E = \dfrac{Lhc}{1000\lambda}$

Wavelength, $\lambda = \dfrac{6.02 \times 10^{23} \times 6.63 \times 10^{-34} \times 3.00 \times 10^8}{1311000}$

$= 91.3 \times 10^{-9}$ m **or** 91·3 nm **or** $9.13 \times 10^{-8}$m
**or** 91nm

**3.** (a) In NO, oxidation state is 2 or +2 or II **or** 2+
In $NO_2$, oxidation state is 4 or +4 or IV **or** 4+

(b)

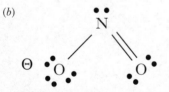

(c) $NO_2^-(aq) + H_2O(l) \rightarrow NO_3^-(aq) + 2H^+(aq) + 2e^-$

**4.** (a) (i) $Fe^{3+}$ $1s^2 2s^2 2p^6 3s^2 3p^6 3d^5$
(ii) $Mn^{3+}$ $1s^2 2s^2 2p^6 3s^2 3p^6 3d^4$
(iii) $Fe^{3+}$ has half filled d-subshell
**or** All d-orbitals half filled in $Fe^{3+}$

(b) Moles of $FeTiO_3$ = 3250/151·7 = **21·42**
Mass of $TiO_2$ = n × FM = 21·42 × 79·9 = 1711 g = **1·71 kg**

(c) $(NH_4)_2[Cu(Cl)_4]$

**5.** (a) Step 4: Rinse beaker with deionised water, add washings to standard flask.
Step 5: Add deionised water up to mark on standard flask.

(b) (i) Murexide or ammonium purpurate
(ii) Octahedral

(iii) Average titre = 23·55 cm³

No of moles of Ni in 100 cm³ solution
$= 0.02355 \times 0.110 \times 4 = 0.0104$

% mass of Ni $= \dfrac{0.0104 \times 58.7}{2.656} \times 100 = \mathbf{22.98\%}$

**6.** (a) T = 300 − 310 K

(b) $\Delta H° = 380 - 420$ (kJ mol⁻¹)

(c) Gradient of line = −1·3 (kJ K⁻¹ mol⁻¹)
or $\Delta S° = 1.22$ to
1·40 kJ K⁻¹ mol⁻¹

$\Delta S° = (+) 1220$ to 1400 (J K⁻¹ mol⁻¹)

**7.** (a) Third order/3$^{rd}$/3

(b) Reaction 3
Rate is independent of concentration of reactants

**or** rate is independent of concentration of ammonia

**or** Concentration of reactant has no effect on rate

(c) $k = \dfrac{\text{Rate}}{[NO]^2[Cl_2]} = \dfrac{1.43 \times 10^{-6}}{(0.250)^2(0.250)}$

$= \mathbf{9.15 \times 10^{-5} \ l^2 \ mol^{-2} \ s^{-1}}$

**8.** (a) $H_2O_2(aq) + 2H^+(aq) + 2Br^-(aq) \rightarrow Br_2(l) + 2H_2O(l)$

(b) $\Delta G° = -nFE°$
$= -2 \times 96500 \times 0.70$
$= -135.1$ kJ mol⁻¹

**9.** (a) A solution in which the pH remains (approximately) constant when small amounts of acid, alkali or water are added, or a solution which resists pH changes when acid/alkali added

(b) Sodium propanoate or potassium propanoate

(c) [salt] $= \dfrac{0.15/80.0}{0.1} = 0.131$ mol l⁻¹

pH $= 14 - 4.76 + \log \dfrac{0.15}{0.131}$

pH $= 14 - 4.76 + 0.059 = \mathbf{9.30}$

**10.** (a) Pharmacaphore

(b)

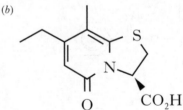

**11.** (a)

| C | H | O |
|---|---|---|
| $\dfrac{50}{12}$ | $\dfrac{5.6}{1}$ | $\dfrac{44.4}{16}$ |
| 4·16 | 5·6 | 2·77 |
| 1·50 | 2·02 | 1 |

**giving $C_3H_4O_2$**

(b) (i) $C_6H_8O_4$

(ii) $2500 - 3500$ (cm⁻¹)
or $1700 - 1725$ cm⁻¹

**12.** (a) Condensation

(b)

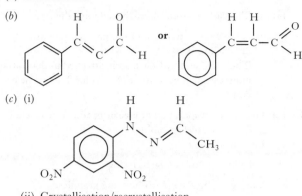

or

(c) (i)

(ii) Crystallisation/recrystallisation

(iii) Measure melting point **and** compare to known data/value

(iv) Yellow **or** orange **or** gold

**13.** (a) (i) Nucleophilic substitution/replacement by a nucleophile

First order **or** unimolecular

(ii)

$$H_3C — \overset{\overset{\textstyle CH_3}{|}}{\underset{\underset{\textstyle CH_3}{|}}{C}}{\large\textbf{+}}$$

(b) (i) Na in ethanol

(ii) Methoxyethane

**14.** (a) 2-hydroxypropanoic acid

(b) Carbon atom ② because it has 4 different groups attached

(c) (i) KCN or NaCN or HCN
**or** correct names

(ii) Hydrolysis/acid hydrolysis

(iii)

$$H — \overset{\overset{\textstyle H}{|}}{\underset{\underset{\textstyle H}{|}}{C}} — \overset{\overset{\textstyle OH}{|}}{\underset{\underset{\textstyle H}{|}}{C}} — \overset{\overset{\textstyle H}{|}}{\underset{\underset{\textstyle H}{|}}{C}} — N\overset{\diagup H}{\diagdown_H}$$

**or** $CH_2CHOHCH_2NH_2$

## CHEMISTRY ADVANCED HIGHER 2012

### SECTION A

| | | | |
|---|---|---|---|
| 1. | A | 21. | C |
| 2. | D | 22. | C |
| 3. | D | 23. | B |
| 4. | A | 24. | D |
| 5. | B | 25. | B |
| 6. | B | 26. | A |
| 7. | D | 27. | C |
| 8. | C | 28. | D |
| 9. | C | 29. | A |
| 10. | A | 30. | B |
| 11 | C | 31. | B |
| 12. | D | 32. | A |
| 13. | D | 33. | B |
| 14. | B | 34. | D |
| 15. | A | 35. | B |
| 16. | A | 36. | B |
| 17. | A | 37. | D |
| 18. | C | 38. | A |
| 19. | C | 39. | D |
| 20. | D | 40. | D |

### SECTION B

**1.** (a) (i) $E = \dfrac{Lhc}{\lambda}$

$$= \frac{(6\cdot02 \times 10^{23}) \times (6\cdot63 \times 10^{-34}) \times (3\cdot00 \times 10^{8})}{1000 \times 405 \times 10^{-9}}$$

$$= 296 \ (kJ \ mol^{-1})$$

(ii) $1s^2 \ 2s^2 2p^6 \ 3s^2 3p^6 3d^{10}$

(b) Photovoltaic effect

(c) Positive holes

**2.** (a) (i) 5 **or** V **or** (V)

(ii) 3 **or** III **or** (III)

(b) $2HNO_2 + 4H^+ + 4e^- \rightarrow H_2N_2O_2 + 2H_2O$

(c) Dicyanocuprate(I)

**3.** (a) The relative radii of the ions.
The relative size of the ions.
The radius ratio of the ions.

(b) Each sodium ion has six chloride ions surrounding it and each chloride ion has six sodium ions surrounding it.

(c) Potassium fluoride

(d) (i) lattice enthalpy and hydration enthalpies (of the ions)

(ii) The entropy change is positive/positive entropy/ increase in entropy/increase in disorder / ΔS positive

**4.** (a) Trigonal planar

(b) Oxygen donates a lone pair of electrons to boron
**or**
correct idea that both electrons have come from the oxygen

(c) Cyclic ethers or ether(s) or furans or cycloethers

**5.** (a) No of moles thiosulphate = $15.25 \times 0.102/1000$
$= 1.56 \times 10^{-3}$    so moles   $Cu^{2+} = 1.56 \times 10^{-3}$

Mass Cu per sample = $63.5 \times 1.56 \times 10^{-3} = 9.88 \times 10^{-2}$

Mass of Cu in key = $9.88 \times 10^{-2} \times 1000/25 = $ **3.95g**

(b) Use distilled/deionised water.
Start with different samples from the key and carry out replicates / duplicates.

(c) EDTA complexes with Cu as well as Ni

**6.** (a) Separating funnel/separation funnel/separatory funnel

(b) 0.127 — 0.130

(c) Different temperature
One of the solutions may be saturated

(d) Ethanol and water are miscible
Ethanol soluble in water
Two layers won't be formed

**7.** (a) $pH = \frac{1}{2}pK_a - \frac{1}{2}\log c$    or    $c = \dfrac{\sqrt{[H^+]^2}}{Ka}$

**or**

$2.5 = \frac{1}{2} \times 4.76 - \frac{1}{2} \log c$    $= \dfrac{\sqrt{(10^{-2.5})^2}}{1.7 \times 10^{-5}}$

$c = \mathbf{5.75 \times 10^{-1}}$ mol l⁻¹    $= \mathbf{0.575}$ mol l⁻¹
(0.575 mol l⁻¹)    (0.588 mol l⁻¹)

(b) Because of the (dark) colour of vinegar or words to that effect, eg the colour change would be hard to see

(c) $CH_3COO^-$ or correct structural formula or $C_2H_3O_2^-$

**8.** (a) "Temperatures **below** 2000 K"
**or** 0 – 2000 K

(b) Slope of line is- ΔS
**or**
$2C + O_2 \rightarrow 2CO$ has increase in entropy
**or**
1 mole gas makes two moles gas
**or**
increase in disorder
**or**
ΔS is positive

(c) Boiling point of Mg/Change of state/magnesium becomes a gas

**9.** $\Delta G = -nFE°$ **or** $E° = \dfrac{-\Delta G}{nF}$

$\dfrac{279.8 \times 10^3}{2 \times 96,500}$

$= \mathbf{1.45}$ V(olts)

**10.** (a) Zero **or** 0

(b) First **or** 1

(c) $s^{-1}$

**11.** (a) (i) Hydrogen or $H_2$

(ii) Hydrogen chloride or HCl

(b) (i) acid chloride, acyl chloride

(ii)

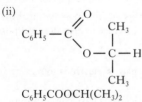

$C_6H_5COOCH(CH_3)_2$

(iii) Faster reaction/ More vigorous reaction/greater yield/needs no catalyst/no equilibrium reached (in an open system)

**12.** (a) $C_8H_{14}S_2O_2$

(b) (i) The carbon atom where the tail joins the ring.

(ii) It has four different atoms or groups (substituents) attached to it.
The "tail" and hydrogen atom attached to that carbon atom can each be in front of or behind the plane of the ring.

**13.** (a) To prevent evaporation or idea of products or reactant or gases or chemicals escaping
To reduce smell.

(b) (Alkaline) hydrolysis/hydrolysing

(c) A few glass beads or anti-bumping granules.

(d) The oily layer disappears/no longer two layers/goes clear/no more oily droplets/cloudy to colourless

(e) $C_6H_5COOC_2H_5$ (150 g) $\rightarrow C_6H_5COOH$ (122 g)
5.64 g $\rightarrow$ **4.59 g**
73.2 % of 4.59 g = **3.36** g
**or**
Using mol calculation, then get 0.0275 mol
Final answer = 3.36 g (as before)

**14.** (a) (i) Electrophilic substitution

(ii) Aminobenzene/Phenylamine/Aniline
**or** correct (structural) formula/$C_6H_5NH_2$

(b) (i)

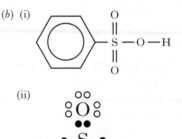

(ii)

**15.** (a) $CH_3CH_2CHClCH_3$

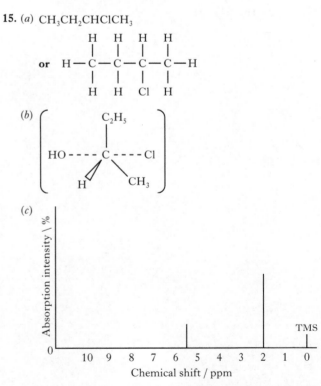

(b)

(c)

Lines can be between 4.5–6.0 and 1.6–2.6 ppm but line of lower chemical shift should be 3× height of other line.

Hey! I've done it

BrightRED
PUBLISHING

Published by Bright Red Publishing Ltd, 6 Stafford Street, Edinburgh, EH3 7AU
Tel: 0131 220 5804, Fax: 0131 220 6710, enquiries: sales@brightredpublishing.co.uk,
www.brightredpublishing.co.uk

Official SQA answers to 978-1-84948-301-8
2008-2012